Understand Philosop S0-AVI-769

Teach® Yourself

Understand Philosophy of Science

Mel Thompson

Hodder Education
338 Euston Road, London NW1 3BH.

Hodder Education is an Hachette UK company

First published in UK 2001 by Hodder Education

First published in US 2001 by The McGraw-Hill Companies, Inc.

This edition published 2012.

Contents

Meet the author

Welcome to *Understand Philosophy of Science*!

I hope I never lose the capacity to be amazed, whether by the dimensions of space, the intricate systems of nature or the workings of the human brain. Much of what we know about the physical world is conveyed to us by science, and so successful has it been in the last four centuries that there is a danger of slipping into an unquestioning acceptance of every latest theory. However, science remains a human activity, and as such it is based on ideas and arguments that should be examined rationally, questioned and sometimes challenged.

Since the first edition was written, more than a decade ago, science has made huge advances, not least in knowledge of the universe, genetics, neuroscience, medicine and information technology. Yet the basic features of the scientific method, and the questions we need to consider in assessing scientific claims, remain largely unchanged.

I hope this book, which aims to give a basic overview of the questions with which the philosophy of science is concerned, will encourage you to develop a critical appreciation of what science has to offer, and the subtleties of its claims.

Mel Thompson

November 2011

In one minute

Science is a massive problem-solving and information-providing enterprise and most people have great respect for what it has achieved. But what does it mean to say that something is 'scientific'? How can one tell valid science from bogus? On what basis can we assess what scientists tell us? How do we know if what we are being told is an absolute truth or merely a temporary theory, adequate for now but soon to be replaced? What is scientific language?

These are just a few of the questions with which the philosophy of science is concerned; many more will come to light as we start to look at the way in which it works, and how it relates to the research projects carried out by scientists.

Science traditionally deals with facts, with information about the world in which we live, gained as hard evidence by means of experiment and analysis. Indeed, the word 'science' comes from the Latin *scientia*, meaning 'knowledge', so science should offer certain knowledge, as opposed to mere opinion. But it is not that simple, and we know that the process of gaining scientific knowledge is one in which the straightforward claim to deal with facts needs to be qualified, both on account of the way we reason from evidence to the framing of scientific theories, and also from the nature of the experiments upon which science is based.

It is generally agreed that scientific theories cannot be conclusively proven, but must be examined in terms of degrees of probability, or appreciated as the best available explanation but not perhaps the only or final one. But, if that is so, what about the knowledge that science has built up over the last four hundred years and the technology that has transformed all our lives? Surely, if something works, the theory on which it is based must be correct! But is such a pragmatic approach the best way to evaluate scientific theories? Should we not at least hope that our theories give us an accurate picture of reality?

We need to examine these and other questions by looking at the way in which science goes about its business.

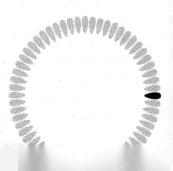

1

Introduction to the philosophy of science

Philosophy is all about asking questions, examining arguments and generally getting to grips with reality. Nobody is likely to get involved with philosophy unless he or she has some sense that the world is an exciting and sometimes confusing place, and that human life is there to be examined as well as enjoyed.

Nowhere is this fascination with the world more evident than in science and the technology that it makes possible. From speculations about the origins of matter, to the understanding and manipulation of genetic information or the workings of the human body, it thrives on the human desire to unlock the mysteries of the world around us – both for the sake of knowledge itself and for the benefits it can offer.

We all know what science is and appreciate what it can contribute to human wellbeing, but why should there be a *philosophy* of science? Surely, science explains itself and validates what it does through the results it achieves. Well, not exactly. For one thing, 'science' is simply a label that is given to certain methods of investigation, and it is quite reasonable to ask whether a field of study is, from the perspective of the rest of science, genuine or bogus. Astronomy is scientific, but what about astrology? Mainstream medicine is scientific, but what of faith healing? Or herbalism? Or homeopathy? And what of the claim that a product is 'scientifically proven' to give health benefits? What does such a claim imply and how can it be verified? These are all questions that require careful attention.

The seventeenth and eighteenth centuries, which saw the rise of modern science, were – generally speaking – a time of optimism, and science was seen in the context of human progress. Reason was hailed as the tool by which humankind would be emancipated from the narrowness of superstition and tradition. The experimental method of the newly developing sciences was a sign of a new commitment to harness reason for the good of humanity. There was a fundamental trust in the human ability to understand and to benefit from that understanding, but, above all, science seemed to offer a degree of certainty about the world.

Knowledge, for science, aims to be proven knowledge, justified by evidence and reason. Nothing is accepted as true unless it has been proved to be so, or there are good reasons to believe that it will be at some point in the future. This reflects the philosophical quest for certainty that goes back to René Descartes (1596–1650), who refused to accept anything that he could not know for certain to be true. He hoped to base all knowledge on self-evident propositions, and thought that reason should take priority over observation. Descartes was aware that his senses frequently misled him. The implication of this – a view which had a long history, prior to the rise of science – was that, if the evidence of our senses did not conform to reason, it was likely that they were in error.

Other philosophers, such as John Locke (1632–1704) and David Hume (1711–76), took sense experience as the basis for knowledge, and it is their approach (known as empiricism) which has provided much of the philosophical underpinning of science.

Although the raw data for science is mediated to us by the senses, we shall be looking at the way in which science has always been at pains to find ways to ensure that our senses are not deceived – in particular, by devising experiments which control nature in such a way that a single feature of it can be checked out, without being too influenced by everything else.

Insight

Life is always too complicated. The only way we can get to grips with it is by simplifying it sufficiently to be able to measure the influence of one thing on another. Science has provided the means by which such measurements and calculations can be made, and, in doing so, has systematized and supported the empiricists' quest for knowledge based on evidence.

Philosophers and scientists

Until the eighteenth century, science and philosophy were not regarded as separate disciplines. **Natural philosophy** was the term used for the branch of philosophy which sought to understand the fundamental structure and nature of the universe, whether by theoretical or experimental methods, and some of the greatest names in philosophy – both before and after science appeared as a separate discipline – were also involved with mathematics and science:

▶ It was Aristotle (384 BCE – 322 BCE) who set out the different sciences and gave both science and philosophy much of its later terminology.

▶ Descartes, Gottfried Leibniz (1646–1716), Blaise Pascal (1623–62) and Bertrand Russell (1872–1970) were all mathematicians. It need hardly be said that science could have made little progress without mathematics, and mathematics is bound up with logic and therefore to philosophy. In their famous book *Principia Mathematica* (1910–13) Bertrand Russell and Alfred North Whitehead argued that mathematics was a development of deductive logic. Thus, much of what is done in science, however specialist in its application, is based on fundamental logical principles.

▶ For some, science was an influence on their overall philosophy and view of the world. Francis Bacon (1561–1626), Locke and others sought to give the scientific method a sound philosophical basis.

▶ Hume, in assessing the evidence of the senses as the basis of knowledge, was influenced by and challenged scientific method.

▶ Thomas Hobbes (1588–1679) saw the whole world as matter in motion – a view to be developed with mathematical precision in Newtonian physics.

▶ Even the philosopher Immanuel Kant (1724–1804), who is generally seen as a writer of abstract and highly conceptual philosophy, wrote *A General Natural History and Theory of the Heavens* (1755) in which he explored the possible origin of the solar system. His distinction between the things we observe (**phenomena**) and things as they are in themselves (**noumena**) is of fundamental importance for understanding the philosophy of science, and especially for defining the relationship between the experiences one has, and the reality which gives rise to such experiences.

As science developed particular forms of experimentation and observation, it naturally started to separate off from the more general and theoretical considerations of philosophy. It also became increasingly difficult for any one person to have a specialist working knowledge of all branches of science, quite apart from all branches of philosophy. Hence the activity of scientists and philosophers started to be distinguished, with the latter carrying out a secondary function of checking on the underlying principles of those engaged in science.

However, it would be wrong to think that the influence has all been one way, with philosophy gently guarding and nurturing its young, scientific offshoot. Science has been so influential in shaping the way we look at the world that it has influenced many aspects of philosophy. The Logical Positivists of the early twentieth century, for example, regarded scientific language as the ideal, and wanted all claims to be judged on the basis of evidence, with words corresponding to external facts that could be checked and shown to be the case. This led them to argue that all metaphysical or moral claims were meaningless. In effect, they were arguing that philosophy should have scientific precision.

Sometimes scientists see themselves as the bastion of reason against superstition and religion. This was a popular view in the eighteenth century, and is still found today, as for example in the work of Richard Dawkins (1941–), who parallels his promotion of science with a criticism of religious beliefs on the grounds that they cannot be justified rationally, effectively using science as a benchmark for proven knowledge.

Insight
The debate between fundamentalist religion and the 'new atheism' is beyond the scope of this book, but can be informed by an appreciation of scientific method and the degrees of certainty it offers.

The role of the philosophy of science

The first example of the philosophy of science, as a separate branch of philosophy, is found in the work of William Whewell (1794–1866), who wrote both on the history of science, and also (in 1840) on *The Philosophy of the Inductive Sciences, Founded upon Their History*.

Generally speaking, the philosophy of science is that branch of philosophy which examines the methods used by science (e.g. the ways in which scientists frame hypotheses and test them against evidence), and the grounds on which scientific claims about the world may be justified. Whereas scientists tend to become more and more specialized in their interests, philosophers generally stand back from the details of particular research programmes, and concentrate on making sense of the overall principles and establishing how they relate together to give an overall view of the world.

There are two problems here:

1. Science is too vast for any one person to have an up-to-date, specialist knowledge of every branch. Hence the philosopher of science, being a generalist, is always going to have a problem with doing justice to the latest scientific work upon which he or she needs to comment.

2. The role of science can sometimes be overstated, with its exponents slipping into **scientism**. Scientism is the view that the scientific description of reality is the only truth there is. With the advance of science, there has been a tendency to slip into scientism, and assume that any factual claim can be authenticated if and only if the term 'scientific' can correctly be ascribed to it. The consequence is that non-scientific approaches to reality – and that can include all the arts, religion, and personal, emotional and value-laden ways of encountering the world – may become labelled as merely subjective, and therefore of little account in terms of describing the way the world is. The philosophy of science seeks to avoid crude scientism and get a balanced view on what the scientific method can and cannot achieve.

The key feature of much philosophy of science concerns the nature of scientific theories – how it is that we can move from observation of natural phenomena to producing general statements about the world. And, of course, the crucial questions here concern the criteria by which one can say that a theory is correct, how one can judge between different theories that purport to explain the same phenomenon, and how theories develop and change as science progresses.

And once we start to look at theories, we are dealing with all the usual philosophical problems of language, of what we can know and of how we can know it. Thus, the philosophy of science relates to three other major concerns of philosophy:

▶ **metaphysics** (the attempt to describe the general structures of reality – and whether or not it is possible to do so)
▶ **epistemology** (the theory of knowledge, and how that knowledge can be shown to be true)
▶ **language** (the nature of scientific claims, the logic by which they are formulated and whether such language is to be taken literally).

This is not to suggest that the philosophy of science should act as some kind of intellectual policeman; simply that it should play an active part in assisting science to clarify what it does. But there are two key questions here:

1 Are there aspects of reality with which science cannot deal, but philosophy can?
2 If philosophy and science deal with the same subject matter, in what way does philosophy add to what science is able to tell us?

The situation is rather more problematic, for there are three different ways (at least) in which we can think of the relationship between philosophy and science:

1 Science gives information about the world, while philosophy deals with arguments and meanings. Philosophy should therefore restrict itself to the role of clarifying the language science uses to make its claims, checking the logic by which those claims are justified, and exploring the implications of the scientific enterprise. It should keep well away from the subject matter within which science deals. This has been a widely held view, and it gives philosophy and science very different roles.
2 However, it may be difficult to draw a clear distinction between statements about fact and statements about meaning. Science does not simply report facts; it seeks to provide theories to explain them. Like philosophy, it is concerned with arguments and the validity of evidence. It uses concepts, and these may need to be revised or explained in different ways. Hence we should not expect to draw a clear line between the activity of science and that of philosophy. This view reflects ideas published in 1951 by the American philosopher W. V. Quine in an important article entitled 'The Two Dogmas of Empiricism' (see Chapter 4).

3 Philosophy describes reality and can come to non-scientific conclusions about the way the world is. These conclusions may not depend on science, but are equally valid. (This reflects an approach taken by philosophers who are particularly concerned with the nature of language and how it relates to experience and empirical data, including Moore, Wittgenstein, Austin, Strawson and Searle.)

And then, of course, one could go on to ask if you can actually do science without having some sort of philosophy. Is physics possible without metaphysics, language or logic? What about all the concepts and presuppositions that scientists use to explain what they find?

We shall see later that science can never be absolutely 'pure'. It can never claim to be totally free from the influences of thought, language and culture within which it takes place. In fact, science cannot even be free from economic and political structures. If a scientist wants funding for his or her research, it is necessary to show that it has some value, that it is in response to some need, or that it may potentially give economic benefit. Scientific findings are seldom unambiguous; and those who fund research do so with specific questions and goals in mind; goals that can influence the way that research is conducted or its results presented.

Insight

For some time, during the middle years of the twentieth century, it was assumed that the principal – indeed the only – role of philosophy was clarification. Since then there has been a broadening out of its function. You may want to consider, as various arguments are presented in this book, whether philosophy has contributed directly to human knowledge, or simply clarified and systematized knowledge that has its source in scientific research or common human experience.

What this book examines

Clearly, there is a huge literature on the philosophy of science, and even more on science itself and its history. This book claims to do no more than touch on some of the key issues, in order to give you an overall perspective on what the philosophy of science is about and – hopefully – to whet your appetite to find out more.

These are the topics we will be exploring in the chapters that follow:

- Since one cannot appreciate what science does without some overall sense of what it has done to date, we start with a brief overview of the **history of science in the West**. Of particular interest here are the logical assumptions made by scientists.
- Science is generally defined by its **method**. We shall therefore examine this, particularly in the context of the rise of modern science from the seventeenth century.
- Science develops **theories**, and the debate about how these are validated or refuted and replaced is a central concern for the philosophy of science. We shall therefore examine the status of scientific claims.
- However, is science able to give us a true and accurate picture of the world? We shall look at claims concerning **scientific realism**.
- It has long been recognized that our observations are influenced by our theories, which suggests that the truths they yield may be relative rather than absolute. We may also assess scientific theories on a pragmatic basis. Are they useful in predicting things we need to know? Are they relevant? Can you accept two apparently contradictory theories at the same time? **Relativism** and **relevance** are key questions to explore here.
- Chaos theory, complexity theory, issues concerning **predictability** and **probability**, whether everything is determined or happens by chance, whether chance may be loaded to produce a particular result – these form a fascinating area of study, raising questions with which philosophy needs to get to grips.
- The philosophy of biology examines particular questions related to **evolution**, the way in which species are related to one another, and the genetic basic of life.
- The scientific quest has never been limited to matters terrestrial. Astronomy was an important feature of the rise of modern science, and today some of the most exciting developments in physics concern **the nature and origin of the universe**.
- Science has much to say about **human beings**, including social and psychological theories that seek to explain behaviour, and particularly – through neuroscience – the way in which our thoughts and experience are related to what happens in the brain. We need to assess the impact of science on our self-understanding and personal meaning, both in this final chapter and throughout the book.

KEEP IN MIND...

1 The philosophy of science examines the methods and arguments used in science.

2 Until the eighteenth century science was known as natural philosophy.

3 The rise of modern science paralleled philosophical questions about the basis of knowledge and a quest for certainty.

4 Some of the greatest philosophers were also involved in science and/or mathematics.

5 William Whewell was the first to write about the philosophy of science (in 1840).

6 Scientism is the view that science is the only valid source of factual knowledge.

7 Philosophy can clarify the language used by science.

8 Philosophy can also examine the logic by which theories are developed.

9 Philosophy can explore the criteria by which we assess competing scientific claims.

10 This book can offer no more than an outline of some of the key issues.

2

The history of science

In this chapter you will:
- *look at some key figures in the history of science*
- *explore the contribution of different periods*
- *consider the central philosophical questions.*

Although the philosophy of science and the history of science are quite separate, it is difficult to see how one could appreciate the former without some knowledge of the latter. There are two key reasons for this:

1 Advances made in science reflect the general ideas and understanding of reality of the period in which they are made and, at the same time, help to shape those ideas. It is therefore interesting to explore the way in which philosophy and science influence one another in an historical context.

2 In order to understand the principles that operate within science, one must know something of the way in which scientists go about their work and why particular questions are of interest to them. They often respond to situations where a previously accepted theory is found to be wanting, and seek to refine or replace it. Thus, looking at the history of science gives us an overview of the questions that science has thrown up in each period.

It is also useful, in order to get the issues that face the philosophy of science into perspective, to have a brief overview of the way in which thinking about the natural world has changed in the West over the last 2,500 years. We shall see that there have been two quite drastic changes in perspective:

1 The first of these took place as the world-view initiated by the ancient Greeks (especially Aristotle) gave way to what was to

become the world of Newtonian physics in what we generally see as 'the rise of modern science'.

2 The second took place as that Newtonian view gave way to the expanding horizons in physics, brought about by relativity, quantum theory, and the impact of genetics on biology, so that, by the end of the twentieth century, the world of science was as different from that of the nineteenth, as the Newtonian world was from the ideas of the ancient Greek or medieval thinkers.

Early Greek thinkers

Ancient Greek philosophy is dominated by the work of Socrates, Plato and Aristotle, but before them there were a group of thinkers generally known as the 'Pre-Socratics', who developed theories to explain the nature of things, based on their observation of the natural world; they were, in effect, the first Western scientists.

PRE-SOCRATIC THINKERS

Thales (sixth century BCE), who is generally regarded as the first philosopher and scientist, considered different substances, solid or liquid, and came to the view (extraordinary for his times) that they were all ultimately reducible to a single element. He mistakenly thought that this fundamental element was water. His answer may have been wrong, but it must have taken a fantastic leap of intellect and intuition to ask that sort of question for the very first time.

Insight

Only one oxygen atom separates Thales from modern physics, for we now consider all substances to be ultimately derived from hydrogen. But what links Thales to modern physics above all is his quest for a single, underlying substance or principle to account for the diversity of physical things.

From the same century, another thinker anticipated later scientific thought. **Heraclitus** argued that everything is in a state of flux, and that even those things that appear to be permanent are in fact subject to a process of change. Our concepts 'river' or 'tree' may suggest something that is permanent, but any actual river or tree is never static. Heraclitus is best known for his claim that 'you cannot step into the same river twice'.

We live now in a world where most people take it for granted that everything from galaxies, stars and species to the cells that make up our bodies are in a constant process of change and development. But Heraclitus came to this view by observation and logic, while those around him saw the created order as static.

Leucippus and **Democritus** (from the fifth century BCE) developed the theory (**atomism**) that all matter was comprised of very small particles separated by empty space. If substances had different characteristics, it was because they were composed of different mixtures of atoms.

Notice the logic used by these early thinkers. They saw that a substance can take on different forms – solid, liquid or gas, depending on temperature (as when water boils or freezes) – and deduced the general principle that the same atoms combine differently at different temperatures. In observing the world, looking for explanations for what they saw, and moving from these to formulate general theories, these early philosophers were doing what we today recognize as science. They had intuition in plenty; what they clearly lacked was any systematic or experimental method.

PLATO

Whereas the Pre-Socratics were happy to study and theorize about the world of experience, there was another side to Greek thought that looked away from immediate experience to contemplate ideal or general concepts. This approach is traced back to Plato (424/423 BCE – 348/347 BCE), who argued that the things we see and experience around us are merely copies of eternal but unseen realities. In other words, I only know that this creature before me is a dog because I have a general concept of 'dogginess'.

Therefore, in order to understand the world, a person has to look beyond the particular things that can be experienced, to an eternal realm of 'Forms'. In his famous analogy of the cave (in *The Republic*), shackled prisoners are able to see no more than fleeting images, shadows cast upon the back wall of a cave by an unseen fire. Only the philosopher turns to see the objects themselves, the fire that casts their shadows and then, beyond the mouth of the cave, to the light of the sun. Reality is thus understood only by turning away from the wall and its shadows – in other words the ordinary world of our experience – and contemplating general principles and concepts.

ARISTOTLE

Aristotle (384 BCE –322 BCE) argued that knowledge of the world comes through experience interpreted by reason; you need to examine phenomena, not turn away from them. The process of scientific thinking therefore owes more to Aristotle than to Plato. He saw knowledge as something that develops out of our structured perception and experience, bringing together all the information that comes to us from our senses.

Aristotle set out the different branches of science, and divided up living things into their various species and genera – a process of classification that became a major feature of science. He was clearly fascinated by all aspects of nature and the way in which things worked together in order to produce or sustain life. Many of the terms we use in science today come from Aristotle.

Most significantly, Aristotle argued that a thing had four different causes:

1 Its **material** cause is the physical substance out of which it is made.
2 Its **formal** cause is its nature, shape or design – that which distinguishes a statue from the material block or marble from which it has been sculpted.
3 Its **efficient** cause is that which brought it about – our usual sense of the word 'cause'.
4 Its **final** cause is its purpose or intention.

For Aristotle, all four causes were needed for a full description of an object. It was not enough to say how it worked and what it was made of, but one needed also to give it some purpose or meaning – not just 'What is it?' or 'What is it like?' but also 'What brought it about?' and 'What is it for?'

There are two important things to recognize in terms of Aristotle and the development of science:

1 His success and importance gave him such authority that it was very difficult to challenge his views, as we see happening later in connection with the work of Copernicus and Galileo.

2 The development of modern science, from about the seventeenth century onwards, was based on a view of the nature of reality in which efficient causation dominated over final causation. In other words, the Aristotelian world, where things happened in order to achieve a goal, was replaced by one in which they happened because they were part of a mechanism that determined their every move. This is something of a caricature (both of Aristotle and of seventeenth-century science), but the shift is clearly important for the philosophy of science.

ARCHIMEDES

The ancient Greeks considered so far are best known now as philosophers rather than scientists (although that distinction was not made in their day), but there is one person who stands alongside them as hugely important for the development of science; that person is Archimedes (287 BCE – 213 BCE).

He is perhaps best known for his bath, from which he leapt shouting 'Eureka! Eureka!' But to reflect on the significance of what he had found, let us reflect on the degree of sophistication of his answer to the problem set. His task was to find if a crown was made of pure gold or if it had been debased. Its weight was equal to the gold supplied. He therefore wanted to measure its volume, and check it against the volume of gold of the same weight, but clearly he could not take the crown apart or melt it down in order to do so. By observing how water was displaced as he got into his bath, he had a simple method of measuring volume – by measuring the volume of water displaced when the crown was immersed in a container full of water. He checked it against the displacement produced by an equal weight of gold, found that its volume was greater, and therefore concluded that a lighter metal had been added.

Notice what was involved in solving this problem:

▶ He has to recognize that the density of a pure substance is the same wherever it is found, and therefore that a change of density will indicate that a lighter substance has been added.
▶ He knows that density is proportional to weight and volume.
▶ He knows the weight; therefore all he needs to do is to check the volume.
▶ The displacement of water provides a practical answer to that task.

You have here a combination of abstract reasoning about substances and their density with a very practical form of measurement. It was a remarkable achievement for Archimedes, but not for the goldsmith, who was executed for having debased the gold in the crown!

Another aspect of Archimedes' work is summed up in his well-known saying that, given a lever and a secure place upon which to rest it, he could move the world. This referred to his work with levers, cranes and pulleys, much of which he developed for military purposes. He produced catapults as well as devices for grappling and hauling ships out of the water. He even used lenses to focus the sun's rays on the besieging Roman ships, causing fires.

Insight

Whereas the pre-Socratics had speculated about the fundamental nature of things, and Aristotle had developed key concepts and systematized the sciences, it is Archimedes who stands out as the practical scientist, using experiment and theory to solve practical problems.

The medieval world-view

The rise of modern science in the seventeenth and eighteenth centuries is often contrasted with the medieval world that preceded it. The caricature is that the medieval world was one based on authority and superstition, whereas, from the seventeenth century, all was based on evidence and reason. However, it would be a mistake to underestimate the way in which medieval thinking, and the universities that developed and disseminated it, made later science possible. But first we need to take a brief look at how Greek thought developed after the time of Aristotle.

AFTER ARISTOTLE

After Aristotle, there developed two very different approaches, both of which find echoes in later science. On the one hand, the **Stoics** saw the universe as essentially controlled by a fundamental principle of reason (the **Logos**). Their aim was to align human reason and behaviour with that overall, rational scheme. This view of the universe, along with their moral principles that followed from it, found parallels in the development of Christianity, in which the world was thought to be ruled by a rational deity, and where Christ was described as the Logos made flesh. On the other hand, the **Epicureans** developed the earlier work of the atomists. For them, the universe was essentially impersonal and determined by the material combination of atoms. The human observer was therefore liberated from any universal principle or authority and free to set his or her own goals.

Insight

Does human reason reflect a universal principle of reason, or does it stand alone in an impersonal, random universe? That is the stark contrast presented by the Stoics and Epicureans. Modern thought has elements of both, since we assume that the universe can be understood rationally, while accepting that its laws work in an impersonal way.

The other major influence on thinking about the nature of the world came with the rise of Christianity, which from a small Jewish sect became an established religion within the Roman empire. Its theology developed against a background of the Greek as well as Jewish ideas, and it naturally tended to find an affinity with ideas from Plato and the Stoics, rather than with those of Aristotle and the Epicureans. As a result, the attitude to the physical world, expressed by Christian thinkers such as St Augustine, tended towards the Platonic view that what happened on earth was but a pale reflection of the perfection of heaven.

This was reinforced by the view of the universe expounded in the second century CE by **Ptolemy of Alexandria** (*c.* AD 90 – *c.* AD 168). In his cosmology, the Earth was surrounded by ten glassy spheres on which were fixed the Sun, Moon, stars and planets. The outermost of these was regarded as the abode of God. Each sphere was thought to influence events on the Earth, which led to an interest in astrology. Everything in the spheres above the Moon was perfect and unchanging; everything below it was imperfect and constantly open to influence and change.

With the dominance of Christian theology, Greek philosophy was banned and was effectively lost to the West through what are generally known as the 'Dark Ages'. However, Aristotle and other thinkers had already been translated into Arabic, and their philosophy continued to be examined and developed, along with mathematics, by a succession of Muslim thinkers. It was only in the thirteenth century, particularly as a result of the translation into Latin of Averroes' commentaries on Aristotle, that his thought was reintroduced into the West and spread through the newly developing network of universities.

THE MEDIEVAL SYNTHESIS

There was an amazing flowering of philosophy in the thirteenth century, with thinkers such as **Thomas Aquinas** (1225–74), **Duns Scotus** (1266–1308) and **William of Ockham** (*c*.1285–1349). Universities were established throughout Europe, and the 'natural philosophy' taught in them (largely based on the rediscovered works of Aristotle) was an important preparation for later developments in both philosophy and science.

Certain features of Greek (particularly Aristotelian) thought influenced the way medieval thinkers looked at the world around them. All physical things were thought to be made up of four elements: earth, water, air and fire. Each element had its own natural level, to which it sought to move. So, for example, the natural tendency for earth was to sink, water to flow down and fire to rise up, thus explaining motion.

It was taken as an unquestionable fact that the heavens were perfect, and therefore all motion in heaven had to display perfection, and thus be circular; there was no scope for irregularities within the heavenly spheres. This belief caused terrible problems when it came to observing the orbits of the planets and their retrograde motion – somehow what was observed had to be resolved in terms of perfect circles, as we shall see later. Although medieval thinkers were logical, they used deductive logic. In other words, they started with principles and theories (e.g. the heavenly spheres are the realm of perfection; perfect and eternal motion is circular) and then deduced what observations ought to follow. This was in stark contrast to inductive arguments, as used by later science, where evidence is gathered as the basis for framing a theory.

Insight
> The relationship between deductive and inductive arguments may be blurred by the fact that we generally interpret evidence in the light of already established ideas. We sometimes see what we assume we ought to see, and are tempted to reject as flawed evidence to the contrary.

The synthesis within which medieval theologians such as Aquinas worked was one in which the basic teachings of Christianity were wedded to the metaphysics of Aristotle. The result was, in its day, both intellectually and emotionally satisfying, combining the best in philosophy with a religious outlook that gave full expression to Aristotle's 'final causation' – in other words, it assumed that everything had a purpose, established by God.

When the medieval person looked up to the stars and planets, set in their fixed, crystalline spheres, he or she saw meaning and significance, because the Earth was at the centre of the universe, and the life of mankind was the special object of God's concern. A rational universe, established by an 'unmoved mover' (a concept introduced by Aristotle but developed by Aquinas), protects the human mind against the despair and nihilism of a world where everything is impersonal or a product of chance.

Insight
> The medieval synthesis of philosophy and theology offered an intellectual structure in which human life – along with everything else in the world – had both purpose and meaning. A synthesis of that sort was not one to be given up lightly, for it offered both existential comfort and (within its own terms) intellectual integrity.

This is not to deny that, at a popular level, life in medieval times was full of irrational beliefs and superstitions. But the arguments between those who held traditional beliefs that were steeped in this medieval synthesis and those who were presenting what became modern science were not all one-sided. It was not superstition versus reason, but the effort to break out of the structures of Aristotelian philosophy, whose very success had led to the assumption that it was infallible.

Aristotle had always emphasized the importance of evidence, and yet the authority given to his natural philosophy could sometimes be given priority over new evidence. Thus **Copernicus** (1473–1543), who considered the usefulness of thinking of the Sun rather than the Earth as the centre of the known universe, and later **Galileo** (1564–1642), who compared the Copernican view with that of Aristotle and Ptolemy, both sheltered their radical views from criticism by claiming that the new view of the cosmos was chiefly to be used as a theoretical model for simplifying calculations, rather than as a picture of what was actually the case (reasonably, since they had no evidence to go on, other than the simplification their Sun-centred model allowed). In later departing from this position, Galileo was seen to be attacking the authority of the traditional Ptolemaic view, which appeared to be confirmed by the Bible.

However, the medieval world was certainly not devoid of imaginative thinkers prepared to explore science and technology in a way that was free from the weight of traditional thinking. **Roger Bacon** (1220–92) based his work on observation and was highly critical of the tendency to accept something as true simply on the grounds of authority. Among many other things, he set down ideas for developing flying machines, while his work on optics led to the invention of the magnifying glass and spectacles.

Leonardo da Vinci (1452–1519) was an amazingly imaginative engineer and visionary, as well as a stunning artist and experienced architect. What you see displayed in his notebooks is the ability to observe nature carefully, and to consider the possible application that such observed mechanisms might have. Like Bacon, he was fascinated by the idea of flying and had notions of planes, helicopters and parachutes.

Generally speaking, medieval thought, following the influence of Aristotle (whose work was taught in universities throughout Europe from about 1250), was based on looking at essences and potentials. Knowing the essence of something revealed its final purpose, and achieving that purpose was to turn its potentiality into actuality. The world was not seen as a random collection of atoms, nor as an impersonal machine, but as the environment where each thing, with its own particular essence, could seek its final purpose and fulfilment. So, for example, the final purpose of the baby was the adult into which it would grow.

With such a philosophy, the task of the person who examines natural things is not physical analysis, but the discovery of essence and purpose, with the conviction that we live in a rationally ordered world.

Insight

With hindsight, this looks like a religious interpretation of the world, but in fact it was simply the outcome of a philosophy that took seriously Aristotle's idea that in describing something one should take into consideration its 'final' cause, not just those effective causes that brought it about. They asked not just 'What is it?' but 'What is it for?'

The rise of modern science

With the Renaissance and the Reformation, there emerged in Europe a general appreciation of the value of human reason and its ability to challenge established ideas. Scepticism was widespread. The seventeenth century witnessed political debate at every level of society, as we see in the English Civil War (1642–51) and its aftermath. The rise of science should therefore be seen against a cultural background of new ideas, of individual liberty, and of the overthrow of traditional authority, both political and religious.

Francis Bacon (1561–1626) initiated what was to become the norm of scientific method, by insisting that all knowledge should be based on evidence and experiment. In doing this, he rejected Aristotle's idea that everything had a final cause or purpose. Rather than observing nature with preconceived notions, he argued that one should start with the observations and reason from them to general principles.

Bacon famously warned against the 'idols' that stood in the way of knowledge, including:

▶ the wish to accept whatever evidence seems to confirm what we already believe
▶ the distortion that results from our habitual ways of thinking
▶ muddles that come from a careless use of language
▶ accepting the authority of a particular person or group.

He also argued that, in gathering evidence, one should not simply try to find cases to confirm one's expectations, but should consider contrary examples as well. In other words, you can't claim that a theory is based on evidence, and then pick and choose which evidence you accept, depending on whether or not it fits the theory! The crucial test of a theory comes when you find a piece of evidence that goes against it. When that happens, you know one of two things: either something is wrong with the new evidence, or else the theory needs to be modified to take this new data into account.

Insight

In this, Bacon anticipated the work of the twentieth-century philosopher Karl Popper, who made falsification the key to progress in science. We shall look at issues connected with the 'inductive' approach in Chapter 3 and Popper's work in Chapter 4.

Overall, it is important to recognize that Bacon saw mechanical causation throughout nature – in other words, that everything happens because of prior causes and conditions (what Aristotle would have called 'efficient causality'). This effectively ruled out Aristotle's 'final cause'; things happened for reasons that lay in the immediate past, not goals that lay in the future.

From an Aristotelian perspective, the essence of an acorn is to grow into an oak tree. Its growth is therefore understood in terms of its efforts to actualize that potential.

From a modern scientific perspective, however, an acorn grows into an oak tree if, and only if, it is in an environment that can nourish it, and it grows on the basis of a genetic code which gives it programmed instructions for how to do it. In effect, Aristotle was saying 'if you want to understand something, look where it's going', whereas science is saying 'if you want to understand something, look what has produced it.'

In many ways, this freedom from always looking for a final goal and purpose enabled science to make progress, by focusing its attention on antecedent causes and setting out the process of change in a mechanical way. However, it also started to separate science from personal and religious views of the world.

This is not to imply that scientists had no religion; Bacon (along with others, including Newton) tended to speak of the two books written by God, one of revelation and the other of nature. But it had the effect of conveniently distancing scientific method from religious and personal views.

For whose benefit?

Should scientific research and technological innovation reflect human moral values, or be free to explore whatever it is possible to know or to do? Those who protest against what they see as unnecessary or potentially harmful research may ask 'What is the point? What do we gain by this?' and demand that science justify itself in terms of immediate human benefit. Yet those who justify research often do so on exactly the same grounds. So, for example, some would argue that it is wrong, in principle, to clone a human embryo, while others would point to the potential use of stem cells from cloned embryos in the treatment for serious disease.

The separation of technical questions ('What can we find out' and 'What can we do with the information we find?') and personal/religious ones ('What is worthwhile?', 'What is the point in this exploration?' and 'What does it say about human life as a whole?') does not suggest that either is invalid as a way of engaging with the world. The issue is how we apply conflicting social and moral values. If everyone agreed on what was worthwhile, there would be no problem.

The early scientists, very much in the spirit of the Renaissance, believed that all progress in human knowledge was in the long run for the benefit of humankind. It was an optimistic period – leading, as they saw it, to a better future, freed from superstition and misery.

COPERNICUS AND GALILEO

The change in the use of reason and evidence is well illustrated by the astronomy of the time. **Copernicus** (1473–1543) was a Polish priest, whose views on the nature of the universe were still highly controversial a century after his death. In *De revolutionibus orbium coelestium*

(On the Revolutions of the Heavenly Spheres; 1543) he claimed that the Earth rotated every day and revolved around the Sun once a year. He also noted that there was no stella parallax – in other words that there was no shift in the relative position of the stars when seen from widely separated places on Earth. He reasoned, from this, that the stars must be considerably further away from the Earth than was the Sun.

Clearly, such findings conflicted with the generally accepted cosmology of Ptolemy. When the book was published, it was given a preface suggesting that it did not claim to represent the way things actually were, but only an alternative view which provided a convenient way to calculate planetary motion. However, the work did establish that, on the basis of carefully gathered evidence, it was possible to put forward a theory that contradicted the previously accepted view of the universe.

What Copernicus actually offered was a better explanation of the retrograde motion of the planets, a motion which had to be accounted for by using a complex system of epicycles (an epicycle being the path traced by a point on the circumference of one circle as that circle rolls round another one) on the older Ptolemaic system. Nevertheless, there were many problems with a Sun-centred model for the universe: if the planets were to move in circular orbits, it is debatable whether Copernicus (who also used epicycles in his account) actually simplified the calculations that much.

Insight

Notice the basis upon which Copernicus put forward his theory. Theories aim at explaining phenomena, and – by the principle known as 'Ockham's Razor' – a simpler theory is to be preferred to a more complex one. Hence, although Copernicus had no solid evidence for his alternative view, he could present his hypothesis as one that was more helpful, even if he could not show at the time that it was any more factually correct than the one it replaced. The fact that the Earth actually moved round the Sun, rather than vice versa, was not proved until 1728, or the rotation of the Earth until 1851.

Copernicus' view was challenged on the basis of common experience. It was assumed that, if the Earth were rotating, we should be flung off by its motion, and yet there was no evidence on the surface of the Earth that suggested such movement. In replying to this, Copernicus' answers were still very Aristotelian. He claimed that evil effects could not follow from a natural movement, and that the Earth's motion did not cause a constant wind because the atmosphere contained 'earthiness' (one of Aristotle's four elements) and therefore revolved in sympathy with the Earth itself. Later, Newton was to explain such things through gravity and the laws of motion, but Copernicus had not made the leap into that new form of scientific thinking.

Insight

Notice the logic here. You put up an hypothesis and, as a method of testing it, ask: 'What evidence should follow if this hypothesis is correct?' Sometimes you may lack evidence; sometimes (as here) you make the wrong assumptions about what will follow – we are not actually flung off a rotating Earth, for reasons that were not known in Copernicus' day.

Hence, it was certainly not the case that Copernicus (and, later, Galileo) stood on one side of a divide, with reason and evidence on their side and Aristotelian tradition and religious bigotry on the other. There was a real dilemma about the evidence and its interpretation. In many ways, it could have been argued that what Copernicus had produced was indeed a view of the universe which simplified calculations, but did not reflect reality. It was not simple prejudice that kept Copernicus' theory from acceptance for more than a century, but some real problems that were not resolved, and the fact that, without a telescope to make his own observations, Copernicus was reliant on naked-eye observations and mathematics.

Insight

Copernicus struggled with the complexity of predicting planetary motion and wondered if there might be some alternative way of explaining his observations. And herein lies his importance for the philosophy of science, for he recognized that there might be different interpretations of the same evidence, and that it was possible to present two alternative theories and ask which of them was the more useful, which was the simpler, and which enabled one to make the best set of predictions. Copernicus therefore represents a crucial step in a direction which, over the next four centuries, was to transform our way of thinking.

Tycho Brahe (1546–1601) considered that the then-known planets (Mercury, Venus, Mars, Jupiter and Saturn) must move around the Sun, but assumed that the Sun, along with those planets, moved around the Earth. The problem was that such theories were attempts to explain the observation of the movements of the planets and Sun relative to Earth. Galileo (who thought that Brahe's cosmology was wrong, preferring the Copernican system) still thought that there was nothing to choose between the systems in terms of explaining the observations, and therefore hoped to show that the tides would prove that the Earth itself moved, and therefore the inherent superiority of the one system over the other.

Johannes Kepler (1571–1630) also felt that the tides were significant. He rightly held that they were in some way caused by the Moon, but had no idea of how a body could have an influence at a distance, and was therefore left to speak of the Moon's 'affinity with water'. His contemporary Galileo criticized all such language, declaring that it would have been more honest simply to say that we do not know. Galileo thought the tides must be caused by the movement of the Earth (just as water will slosh around within a bucket when it is moved). His view sounded more rational, but was plainly incorrect.

Kepler, meanwhile, marked yet another radical break with Aristotelian thinking. In observing the orbit of Mars, he found a difference between what he observed and what he calculated should be the case. He concluded that the orbit was elliptical rather than circular, with the Sun at one focus of that ellipse. This contradicted the Aristotelian assumption that perfect motion was circular. Previously, astronomers had tried to retain the perfection of heavenly motion by suggesting that the orbits of the planets were in fact epicycles.

Insight

Epicycles were the attempt to make observation fit theory – to get what looked like an elliptical orbit by a system of circles rotating around other circles. With Kepler's observations we can ask 'Why circles?', 'Why should circular motion represent perfection?' and 'Why perfection anyway?' and new possibilities start to open up.

Galileo Galilei (1564–1642), described by Einstein as 'the father of modern physics', sought to demonstrate that nature operates in a regular, mathematical way, by using instruments and conducting experiments to furnish evidence to back up his arguments.

A key feature of his work was the use of the telescope, which he developed from the existing spyglass. He saw that the 'moving stars' (planets) were not like the fixed stars, but were orbs, glowing with reflected light. He also observed the phases of Venus, making it quite impossible to accept the cosmology that had been proposed by Ptolemy, since Venus could be seen to go around the Sun. Without observations enhanced by the telescope, there had been no evidence to decide between different views of planetary motion. The only problem at this point was that, although the phases of Venus proved that the Ptolemaic system was wrong, it could not actually prove that the Copernican alternative was correct. True, the evidence was accounted for more simply with that view, but simplicity did not constitute proof.

Of course, as is well known, the Holy Office declared in 1616 that it was a 'revealed truth' (i.e. found in the Bible) that the Sun moved round the Earth, but Galileo got round this by agreeing with his then-friend Pope Urban VIII to consider Copernicus only as a useful hypotheses for astrological calculations – a reasonable view to take in the absence of absolute proof one way or the other. But in 1632 Galileo published his Italian-language *Dialogue of the Two Chief World Systems* in which he directly compared Copernicus' view with that of Ptolemy, and came to the conclusion that Copernicus was right.

The implication of this work was that Copernicus had described the *actual* universe and not simply offered a useful *hypothesis* for making calculations – thus going against his earlier agreement with the Pope.

The matter was made rather more complex because Galileo wrote in dialogue form, in which two characters present the alternative world systems, and a third tries to judge between them. This enabled him to present the case for Copernicus through the mouth of his character Salviati, without actually saying that he endorsed it himself. Against the charge that, if the Earth moved, one would feel the movement, Galileo argued that no experiment conducted on the Earth could ever prove its movement. He cites the example of a large ship. If one is in an inner cabin of a large ship, there is no sense of motion. Equally, butterflies or fish in such a cabin could move normally, their particular motion quite oblivious of the larger movement of the ship in which they were being carried.

Of course, in the end, it is clear that the dialogue is not evenly balanced; the Copernican side prevails. The movement of the planets, the annual shifting of the path of sunspots, and the tides all suggest that the Earth in fact (not just in theory) moves around the Sun rather than the other way around. And this, of course, then brought Galileo into conflict with the officially agreed position. Galileo was put on trial and forced to recant. He had used reason and observation to challenge the literal interpretation of Scripture and the authority of the Church.

Although it is popular to see this as a significant moment when authority was challenged by scientific evidence and reacted with authoritative high-handedness, it was far from straightforward. Both religious and scientific communities were divided on the issues. It is clear that a significant number of senior churchmen (including Pope Urban VIII) had earlier been supportive of Galileo's work. One prelate had written an apology to him after a priest had criticized him from the pulpit for propounding views that contradicted the literal meaning of Scripture. The prelate was clearly irritated by the naive and literalist approach taken by the priest.

Two important goals for Galileo were rational explanation and simplicity:

▶ **An example of rational explanation:** In his work on projectiles, Galileo worked out theoretically why a 45-degree angle enabled a gunner to achieve the greatest range. This was known from practical experience, of course, and could be reproduced by experiment; but such experimental method was only a means to the end of a rational explanation.

Insight

This, of course, was to become the distinguishing feature of the whole of what we tend to refer to as Newtonian science – the framing of laws by which the actual motions and behaviour of objects can be understood and predicted.

▶ **An example of simplicity:** The movement of sunspots, which he observed carefully, required the Sun to go through a complex set of gyrations in order to explain why the path of the spots was either convex or concave when viewed from the Earth on all but two days in the year. He opted for the simpler explanation that the path of the spots was being observed from a moving and tilting Earth.

If there are particular moments in the history of science in which there are major philosophical breakthroughs, the later work of

Galileo is one of them. In presenting his book on the *Two New Sciences* (being the sciences of materials and of motion) which was published in 1638, he made this astounding statement:

The cause of the acceleration of the motion of falling bodies is not a necessary part of the investigation.

This has important implications. Galileo was examining the actual way in which something happened, not why it happened, thus finally parting company with Aristotle.

Galileo backed up his work by setting up experimental demonstrations. Earlier in his career he had dropped balls of different weights from the top of the leaning tower of Pisa in order to demonstrate that things accelerate downwards at the same speed. (In actual fact, it didn't work quite as he had planned, since added wind resistance meant that the different balls struck the ground at slightly different times – but in any case they landed much more closely in time than would have been predicted by Aristotle, who thought that bodies accelerated towards the Earth in proportion to their weight.)

THE NEWTONIAN WORLD-VIEW

In his key work, *Philosophiae naturalis principia mathematica* (Mathematical Principles of Natural Philosophy; 1687), **Sir Isaac Newton** (1642–1727) examined the world along mathematical principles. With absolute time and space and laws of motion, concepts such as mass, force, velocity and acceleration, he provided a comprehensive framework for the development of physics. His was the defining voice of science until challenged in the twentieth century.

Even now, while we recognize that his physics is inadequate for examining the extremes of scale in the universe, whether cosmic or subatomic, it is the basic laws of motion set down by Newton that serve as the practical guide for the majority of ordinary physical calculations, and which have given rise to most of the technologies that shape our lives.

Whereas Aristotle would have said that an object seeks its natural place within the universe, Newton's first law states that it remains in a state of rest or in uniform motion in a straight line unless acted upon by a force. There is no overall theory of purpose or end, therefore, but only one of forces bringing about change.

The massive contribution of Newton does not rest only on the laws of motion, however radical they proved to be, but in the general view of the universe as a rational and understandable place, whose every operation could be plotted and expressed in mathematical terms. The world of Newton was perhaps (from the standpoint of the twenty-first century) small, crude and mechanical; but it represented a basis upon which – for the next two hundred years – there could be serious developments in both theoretical science and also practical technology.

With the coming of the Newtonian world, philosophy changed its function, from permitting metaphysical speculations about the nature of reality, to examining the logic of the newly formulated principles, and justifying them on the basis of the scientific method used to produce them. So, for example, the philosopher Immanuel Kant, challenged by the success of Newtonian physics, and yet recognizing the limitations of what we can know through sense experience, turned conventional philosophy on its head and recognized that ideas such as space, time and causality were not simply 'out there' in the world, waiting to be discovered, but were fundamental features of the way in which the human mind encounters and makes sense of its experience.

Comment

In this section, we have outlined the popular view that the rise of modern science ousted the work of Aristotle and that the scientists of the sixteenth to eighteenth centuries were battling the authority given to his philosophy by the Catholic Church.

There is limited truth in this caricature. Aristotle (and other ancient philosophers) continued to be studied and were an important influence long after those developments we refer to as 'the rise of modern science'. The key feature of this period (as we see in the work of Bacon and Descartes, for example) is that Aristotle's 'four causes' were effectively reduced to two: 'material' and 'efficient'. The effect of this was to portray the world as a machine, essentially comprised of physical objects causally related to one another. What is lost is Aristotle's 'formal cause' –

that which gives shape and coherence to a complex entity – and his 'final cause', which is its overall purpose and aim.

The concentration on 'efficient causation' enabled great advances to take place in terms of the prediction of physical events, and the framing of scientific laws. Those parts of Aristotle that science no longer took into account sometimes became labelled – unfairly to Aristotle and often pejoratively – as 'metaphysics' and seen as the realm of speculation and the philosophy of religion, in contrast to the knowledge offered by the empirical sciences.

We should not forget the enormous significance of other figures in the history of science – for example, **Robert Boyle** (1627–91), who did fundamental work in chemistry, showing how elements combine to form compounds, or, more than a century later, **John Dalton** (1766–1844), who examined the way atoms combine to form molecules.

There was a steady development of scientific theories and also in the establishment of science: both the Royal Society in England and the Académie des sciences in France were founded in the seventeenth century.

New instruments promoted the careful examination of the world. The telescope was invented in the early years of the seventeenth century, and developed by Galileo and used by him to controversial effect. Microscopes were also being developed, and by the latter part of the century Robert Hooke's best-seller *Micrographia* (1665) was fascinating people by showing images of things previously far too small to be observed. The seventeenth century saw the development of the pendulum clock by Christian Huygens, and by the mid eighteenth John Harrison was perfecting his timepiece to achieve an accurate calculation of longitude, an invaluable aid for those travelling by sea. Twenty years later the Montgolfier brothers had taken their first manned balloon flight (1783), and by the end of the century Count Volta had produced the electric battery.

In the 1660s Robert Boyle set out his forecasts for the future of science, including the art of flying, the cure of diseases at a distance or at least by transplantation, and the use of drugs to ease pain. He hoped that science could aid the recovery of youth, or at least some aspects of it, and even looked to 'the transmutation of species in minerals, animals and vegetables'. These aspirations of 350 years ago set a challenging agenda, but indicate so clearly the enthusiasm of that age for what science might achieve.

Nineteenth-century developments

The changes brought about by science and technology in the nineteenth century were astonishing. The early decades saw the dominance of steam power – in railways, factories, steamships, pumps. But from the 1830s another form of power was to transform technology, electricity – first with the dynamo and motor, and later with the electric telegraph, which when a transatlantic cable became operative from 1866 offered instant international communications. With Alexander Graham Bell's invention of the telephone in 1876, and Guglielmo Marconi's radio transmission in 1895, the world was set for a revolution in personal communication.

With the telephone, telegraph, a postal service, steam railways, factories, buildings constructed using steel, and the arrival of the motor car in 1885, the world had been transformed. By the end of the century, life and health could be improved by taking aspirin and having an X-ray, or could be ended by a machine gun or an electric chair. Although there was a moral and romantic reaction against the 'dark satanic mills' of the industrial revolution, it is difficult to imagine that many towards the end of the nineteenth century could seriously have challenged the overall benefits to humankind offered by science and technology; it had become fundamental to the whole way of life in developed countries. And, of course, in terms of the perceptions of life in general, it seemed to offer humankind the prospect of increasing mastery over its environment.

THE SCIENCES OF HUMANKIND

Probably the greatest single change in human self-understanding to come from the nineteenth century was brought about by the theory of evolution. But alongside this was another, less obvious, but equally important development: the use of statistics. Today we take it for granted that any examination of personal or social life will be set against a background of statistical information. For example, in order to study possible environmental factors in the incidence of disease, one looks at statistics for the disease in various environments or among people who do certain work, or have a particular habit (such as smoking or taking no exercise). On the basis of these, evidence is put forward in the form of 'people who do X are 80 per cent more likely to contract Y'. Thus we often accept statistical correlations as good evidence for one thing causing another, even if the actual mechanism by which that cause operates is unknown. Modern sciences of humankind – psychology, sociology, political science – are quite unthinkable without a foundation of information gathered in the form of statistics. But it was only in the nineteenth century that humankind started to become the object of study in this way.

Moreover, as these statistics were analysed, for example, by the sociologist **Émile Durkheim** (1858–1917), it started to appear that there were trends in human behaviour that could be measured and predicted. Durkheim came to the conclusion that there were social 'laws' at work that could be known statistically, since they produced sufficient pressure on individuals to account for a certain number of them following a particular line of action. Of course, it was not assumed (then or now) that statistics could show laws of the same sort as the laws of physics. There was no way that individual choice could be determined by them. But it was argued that, at the social level and in sufficiently large numbers, behaviour could be mapped and predicted.

Insight

As we shall see later, this has implications for an understanding of personal freedom. Are the individuals who go to make up those statistics really free? Are they forced (at least to some extent) to follow a social trend, even if they are unaware of it? We need to keep in mind that statistical information *describes*, rather than *explains* – which are two very different things.

On the political side, this period also saw the work of **Karl Marx** (1818–83), who, through analysis of historical causes of conflict and relating them to the class structure of society, was able to map out a theory of change and evolution in the political and social arena. Thus, he appeared to offer a 'science' of humankind's behaviour in this sphere. As will be seen later, some twentieth-century philosophers (e.g. Karl Popper) were to criticize Marxism as pseudo-science on the grounds that it did not allow contradictory evidence to determine whether its theories were right or wrong, but simply adapted its interpretation to take all possibilities into account.

Nevertheless, with Marx we do have a theory which purports to use scientific method to study mankind – and, of course, not just to understand the way things are, but to change them. Thus, we find that the attention of science has been turned towards humankind, and human behaviour becomes open to study and scientific analysis.

However, there remained an enormous issue in terms of nineteenth-century science and self-understanding, one that appeared to get at the very core of what it means to be a human being, supreme over lesser species: the theory of evolution by natural selection.

THE CHALLENGE OF EVOLUTION

The work on evolution, leading up to that of **Charles Darwin** (1809–82), took two forms: an examination of fossil evidence, and theories about how species might develop.

William Smith (1769–1839) studied rock strata and the fossils contained in them. He recognized that the deeper and older strata showed life forms different from those found in the present, and concluded that there must have been many successive acts of creation. Geology had emerged as a science capable of revealing history.

The same evidence led **Charles Lyall**, in his book *Principles of Geology* (published between 1830 and 1833), to a different view. He argued for a continuous process of change, rather than separate acts of creation, to account for the differences between the layers of fossils. His view was termed 'uniformitarianism'. What he did not have, of course, was an understanding of the mechanism that could drive such change – but neither did **Robert Chambers** (1802–71) in his controversial book *The Vestiges of the Natural History of Creation*, published anonymously in 1844. Its view that new species could appear challenged both the biblical account of creation but also the sense that humankind might have a unique place within the scheme of things.

Other scientists were already framing the ideas that led more directly to Darwin's understanding of evolution. His grandfather, **Erasmus Darwin** (1731–1802), thought that all organic life formed a single living filament over the earth, and that new species could develop from old. He saw humankind as the culmination of the evolutionary process, but not separate from it. His main book, *Zoonomia* (1794), was mainly a medical textbook, but it included his ideas about evolution. In many ways, his thinking anticipated that of his grandson.

A key figure in the rise of evolutionary theory was **Jean-Baptiste de Lamarck** (1744–1829). He believed that you could categorize species in terms of their complexity, with every species tending to evolve into something more complex. The way in which this happened, according to Lamarck, was through offspring inheriting the characteristics which an individual had developed during its lifetime. In other words, a person who had developed a particular strength or ability would be able to conceive a child who had that same quality, and thus move evolution in that direction. This (generally known as the 'Theory of Acquired Characteristics') became a widely held view during the nineteenth century, to be overtaken by Charles Darwin's alternative explanation in terms of natural selection.

Another scientist who had a profound impact on the emerging idea of evolution was **Thomas Malthus** (1766–1834). He observed that, in any situation where there were limited supplies of food, the populations of species would be limited. Within the species, there would be competition to get such food as was available, as a result of which only those who were strongest, or in some other way best able to get at the food, would survive. These observations, set out in his *Essays on the Principle of Population* (1798), were to provide Charles Darwin with a mechanism he needed to explain the process of evolution.

The breakthrough in the scientific understanding of evolution came with Charles Darwin himself. His *On the Origin of Species* (1859) was controversial because for the first time it presented a theory by which one species could develop from another. His story is well known. He was convinced – by the variety of the species he had seen, especially on the Galapagos Islands in the early 1830s; by the way they were adapted to their surroundings, and by the way in which some living species were related to fossils – that one species must indeed develop out of another. He worked for the next 20 years to develop the theory of how this happened.

Darwin was well aware of the ability of farmers and others to breed particular forms of animals. It was clear that, by selecting particular individuals for breeding, a species could be gradually changed. He also observed, when on the Galapagos Islands, the way in which finches on different islands tended to have different beaks, in relation to the type of food available. He concluded that there had been just a single form of finch originally, but that on each of the islands the isolated communities of finches had each developed in response to those characteristics which gave them an advantage in terms of gathering food. If you needed a short stubby beak for cracking nuts, then – if nuts were your main source of food – that characteristic would spread in the finch population, since the short, stubby beaked ones were more likely to succeed in surviving to breed.

This reinforced what Malthus had said about the control of population numbers through finite food supplies. Variation and the overall limitation of food thus gave Darwin what he needed for his theory about the mechanism of evolutionary change.

In his theory of **natural selection**, Darwin argued:

▶ some individuals within a species have characteristics that help them to survive better than others
▶ those who survive to adulthood are likely to breed, and thus pass on their characteristics to the next generation
▶ thus, with successive generations, there will be an increase within a species of those characteristics which improve its chances of survival
▶ the characteristics of a species are thus gradually modified.

The first four chapters of *On the Origin of Species* outline the process by which his theory is established. He starts with looking at the process of breeding domesticated animals. Then he moves on to consider the variety within species in the wild. He links this with Malthus, by exploring the struggle for existence. Then, from this, he is able to formulate the theory of natural selection. In effect, Darwin was suggesting that the environment within which any species lives carries out, in a natural and mechanical way, what farmers and breeders have long been doing to domestic animals – it has selected favoured characteristics for breeding.

With hindsight, given what had been examined before Darwin, the theory seems obvious, and merely a gathering together of insights that had already been explored by others. In fact, however, it was the clarity of the argument in *On the Origin of Species* that made it so crucial in shifting the whole way of thinking about evolution. What Darwin had produced was a convincing argument about the mechanism by which evolution could take place, a mechanism which was impersonal and certainly required no divine designer to bring it about. By implication, it placed humankind on a level with all the other species, for it too had emerged through a process of natural selection.

Later, Darwin was to go on to explore the implications of this for humankind. His books dealing specifically with human evolution are *The Descent of Man* (1871) and *Expressions of the Emotions in Man and Animals* (1872). But, from the perspective of the history of science, it is *On the Origin of Species* that marks the decisive step.

Insight

Evolution is a theory which is at once elegant and simple, but devastatingly mechanical. In it, Aristotle's 'final causation' can have no place – what appears as design or purpose is but the operation of cumulative chance. Hardly surprising that Darwin's theory proved so controversial.

The genius of Darwin, and why his theory is so important for an appreciation of science, is that he moves from a great deal of data, meticulously gathered over a number of years, through a series of carefully argued steps, to an overall explanation that changes our whole view of the world: science at its best.

No theory is sacrosanct, and even the best change and develop. So, for example, Darwin did not know how the variations between members of a species were produced. Our knowledge of genetics has now shown the random process of errors which occur when genes are copied, some of which may be beneficial – an insight which led Professor Steve Jones, writing about his book *Almost Like a Whale* (Doubleday, 1999), to describe life as 'a series of successful mistakes'. The key thing about a great scientific theory, however, is that it is able to be used as the basis for further work, as well as giving a new and original view of its subject matter. Darwin shifted our understanding of biology away from the idea of distinct species, each with its own fixed essence, to one in which all species are related in a rich and evolving tapestry of life.

Relativity and thermodynamics

Having overthrown the overarching authority of Aristotle in the medieval world-view, science, by the latter part of the nineteenth century, seemed to have become established on absolutely solid foundations, based clearly on reason and evidence. Some (e.g. the philosopher Ernst Haeckel (1834–1919), who in his book *The Riddle of the Universe* (1899), proposed a philosophy of scientific materialism) believed that little still remained to be discovered; Newtonian physics and Darwinian evolution had, between them, provided a secure framework for answering scientific questions, which – by implication – were the only ones worth asking.

All of that, of course, was swept away by the discoveries of the early twentieth century. Once again, authority was challenged, but this time it was the authority of Newtonian physics that came under attack. The philosophy of science suddenly had to open up to the possibility that there may be equally valid but contradictory ways of understanding phenomena. The general perception of science also changed – from what might have been seen as the triumph of common-sense reason and evidence in the seventeenth century, to the acceptance of ideas that seemed far removed from logic and common sense. By the end of the twentieth century, the world has been revealed as a far more confusing and complex place that had been thought a century earlier.

Note

It would be quite impossible to summarize here all the major developments in science that took place during the twentieth century; nor would that be necessary for the purposes of understanding the philosophy of science. What we need to grasp are the key features that distinguish recent developments from the earlier world of Newtonian physics, and the implications they have for understanding the way science goes about its task and justifies its results.

RELATIVITY

The two theories of relativity, developed by **Albert Einstein** (1879–1955) in the early years of the twentieth century, showed the limitations of Newtonian physics:

1 'Special Relativity', which he put forward in 1905, may be summarized in the equation $E = mc^2$. This links mass and energy. E stands for energy, m for mass and c for the speed of light. The equation shows that a small mass is equivalent to a large amount of energy.

2 'General Relativity', which followed in 1916, argued that space, time, mass and energy were all linked. A famous prediction made by this theory was that a strong gravitational field would bend rays of light, and this was confirmed soon afterwards by observing the apparent shift of location of stars during an eclipse of the Sun (see Chapter 3). Both space and time are influenced by gravity. Space is compressed and time speeds up as gravity increases. But gravity is proportional to mass (the larger the mass, the greater its gravitational pull) and, according to the theory of Special Relativity, mass is related to energy.

The implications of all this it to deny a single or definitive perspective. In observing something, the position and movement of the observer has to be taken into account. It might have been adequate in Newton's day to assume a static position on Earth from which things could be observed, but Einstein showed, effectively, that there was no one fixed point; everything is relative.

His theory also had the effect of setting a limit to any known process or connection: the speed of light. If two objects are moving apart at a speed greater than the speed of light, they can have no connection with one another. The speed of light therefore determines the dimensions of the known universe.

THERMODYNAMICS

The fundamental interconnectedness of all physical states, shown by relativity, is illustrated also by the three laws of thermodynamics.

▶ The **First Law of Thermodynamics** states that there is a conservation of mass-energy; the one may be turned into the other, but the sum total of the two remains constant.

▶ The **Second Law of Thermodynamics** is that in every process there is some loss of energy and heat. Therefore, as organized or complex things interact, they gradually give off energy and therefore tend to become cooler and less organized. In other words, everything (and that means the whole universe) is gradually moving in a direction of general disorder or **entropy**.

Note

In looking at this general drift towards entropy, there is a difference between open and closed systems. In a closed system, everything gradually winds down as energy is dissipated. An open system (i.e. one which has the capacity to take into itself energy from outside) can be self-sustaining, and can grow to become more complex. Thus, individual parts of the world can be seen to 'warm up', while the universe as a whole – which by definition must be a closed system – is 'cooling down'.

Your closed-system cup of coffee will gradually cool; the open-system mould on your neglected sandwich will grow and spread.

▶ The **Third Law of Thermodynamics** shows that the cooler something is, the less energy it can produce, and that all energy ceases to be produced at a temperature of −273° Kelvin. This sets a limit to the universe: at absolute zero, everything stops.

In Newtonian physics, because it was concerned with a limited set of conditions as found on Earth, these fundamental limits did not apply. Thermodynamics shows that the universe is not a piece of machinery in perpetual motion; every process is paid for by the dissipation of energy.

The impact of quantum mechanics

The debate about quantum mechanics, particularly that between Einstein and Bohr, conducted in the 1930s, raised fundamental issues about what science can say, and is therefore of great interest to the philosophy of science.

Quantum mechanics developed through an examination of subatomic phenomena, and therefore concerned issues that could not arise before the early twentieth century. The idea that matter was composed of atoms separated from one another by empty space was not new, having been put forward by Leucippus and Democritus in the fifth century BCE (see above). But until the discovery of the electron in 1897 the atom had been thought of as a solid but indivisible speck of physical matter. The atom was then visualized as having a nucleus made up of protons and neutrons, with electrons circling round it, like planets in a solar system. Once it reached that stage, theories were developed about subatomic particles, their behaviour and relationship to one another. Matter was soon to be seen as particles bound together by nuclear forces.

> **Note**
>
> The term 'quantum mechanics' came from the work of **Max Planck (1858–1947)**, who found that radiation (e.g. light or energy) came in small measurable increments or packets ('quanta') rather than as a continuous stream.

When dealing with things that cannot be observed directly, it is difficult to decide if an image, or way of describing them, is adequate or not. You cannot simply point to the actual thing and make a comparison! Hence, when dealing with subatomic particles, all imagery is going to be limited.

A major problem was that particles seemed to change, depending upon how they were observed. Unlike the predictable world of Newtonian physics, quantum theory claimed that you could not predict the action of individual particles. At most, you could describe them in terms of probabilities. Observing large numbers of particles, you could say what percentage were likely to do one thing rather than another, but it was impossible to say that of any one particular particle.

In 1927 **Werner Heisenberg** (1901–76) showed that the more accurately the position of an particle is measured, the more difficult it is to predict its velocity (and vice versa). You can know one or the other, but not both at the same time. But was this 'uncertainty principle' a feature of reality itself, or did it simply reflect the limitation of our ability to observe and measure what was happening at this subatomic level?

This led to considerable debate about the way in which quantum theory should be interpreted. In particular, the issue that divided Einstein and **Niels Bohr** (1885–1962) was whether quantum theory showed what was actually the case, or simply what we could or could not observe to be the case. (Bohr held the former position, but Einstein refused to accept that actual events could be random.)

The issue here is quite fundamental. In traditional mechanics and statistics, we may not be able to know the action of individual things (any more than we can know what an individual voter is going to do at an election), although we can predict general trends. But at the same time, it is believed that each individual is actually determined, although we cannot know all the factors involved, or cannot measure them. Quantum mechanics goes against this, claiming that (at the subatomic level) measurement is fundamentally a matter of probability, and that it is never going to be possible, even in theory, to know the actual behaviour of individual particles.

The debates between Einstein and Bohr are instructive for the philosophy of science, not just in terms of the interpretation of quantum mechanics, but for the method of argument. The debates were conducted largely on the basis of thought experiments, the possible results of which were then checked mathematically to see if they made sense. In other words, a question was posed in the form of a hypothetical experiment (e.g. 'Suppose we had an apparatus that could measure a single photon passing through a hole into a box…') and the implications of that are then teased out logically (e.g. 'Could we then measure its weight? If so, that would show us…'). Such thought experiments enable scientists to explore the logic of a line of investigation, even if the experiments they discuss are not capable of being performed physically. An example of such a thought experiment, called 'Schrödinger's Cat', is outlined in Chapter 5.

Genetics

It is difficult to gauge the full extent of the revolution that has sprung from the discovery of the structure of DNA, made by **Francis Crick** (1916–2004) and **James D. Watson** (1928–) in 1953. It has provided a remarkable way of exploring, relating and (controversially) manipulating living forms. In many ways, the genetic basis for life is the archetypal scientific discovery – the structure that carries the instructions from which all living things are formed. It has revolutionized the biological sciences in the same way that relativity and quantum mechanics have revolutionized physics.

Briefly, deoxyribonucleic acid (DNA) is made up of two strands of chemical units called nucleotides, spiralled into a double helix.

It is found in the nucleus of every living cell. The entire strand of units is called the genome, and along it are sequences (the genes) which give the instructions for building protein molecules, which themselves build living cells. Human DNA is found in 23 pairs of chromosomes in the nucleus of a cell. Thus, the 'genetic' information, contained in the DNA, determines the character of every organism.

This has many implications for the philosophy of biology, and the potential uses of genetics raise questions of a more general metaphysical and ethical nature. As we saw above, Darwin's theory of natural selection depended on the idea that there would always be small variations between individuals of a species, and that those with particular advantages would survive to breed. What Darwin did not know was the mechanism for these random variations. We now know that it is because the genetic code is not always copied exactly, leading to mutations, some of which survive and reproduce. Such mistakes are rare, however, and they can be passed on to offspring only when they occur in particular cells. Nevertheless, genetics has endorsed Darwin's theory, by showing how variations can occur, and thereby giving the raw material upon which 'natural selection' can go to work.

This illustrates an important point for the philosophy of science:

> that a theory should not be rejected simply on the grounds that it depends on phenomena for which there is no present explanation. New discoveries may provide evidence to give retrospective support for theories, as genetics has done for natural selection.

The digital revolution

Computing is not new, but previously it was time consuming; there is a physical limit to what can be done with an abacus! The first steps towards modern computing were taken by **Charles Babbage** (1791–1871) in 1820, who hit upon the idea of devising a mechanical device for mathematical computation. Although his initial efforts were not well received, he persisted, using ideas for inputting data by punch card and other features that were to become part of modern computing.

A real breakthrough in computing came with the work of **Alan Turing** (1912–54). The crucial practical difference between his work and the earlier efforts of Babbage was the use of simple digital technology, and the recognition that all that was required was a binary (off/on) device, such as a telephone relay. When a complex mathematical question was broken down into a sequence of binary choices, a machine could perform each of those operations, and therefore, by the application of logic, could solve problems. The work received a great boost through the efforts to break Nazi codes during World War II. Turing's 'universal machine' contributed to the code breaking at Bletchley Park, which gave the allies invaluable information about enemy intentions.

By the first years of the twenty-first century, children were playing on computers far more powerful than anything used in the NASA space programme that put a man on the Moon thirty years earlier. But alongside that has come the crucial ability to network computers, and through that what may turn out to be the most significant of all technological developments – the Internet. There can be no doubt that life is perceived quite differently by those who routinely use it as a source of instant information from any part of the globe, and the ability to communicate, contribute and relate to people and institutions anywhere gives a whole new social dimension to technology.

Today we create and share instantly – our laptops, tablets and smartphones keep us in touch with an invisible network that is sophisticated both technically and socially. And with the arrival of cloud computing, our personal information is stored remotely, to be accessed by us at will from any point on the globe and by using a whole range of devices – features unthinkable only a generation ago. When Steve Jobs, co-founder and CEO of Apple, died in October 2011, obituaries and tributes flowed in from all parts of the globe, and millions of people were able to share their thoughts about him online through social media sites. He was hailed as a man who changed the way we see the world. That in itself says a great deal about the way in which the digital revolution has impacted on the global community.

The development of personal and online computing is perhaps the most immediately obvious aspect of the digital revolution, but the issue – certainly as far as the philosophy of science is concerned – goes much deeper. Since the time of the ancient Greeks, mathematics – as a tool of logic and calculation – has been important for the development of science, and that continues to be the case. And that has continued to be the case. From cosmology to nuclear physics, mathematics is a fundamental tool of scientific work. Then, as we saw, from the nineteenth century, statistics started to be gathered and analysed. This gave social scientists and others a way of making causal connections that had the backing of large numbers of observations, expressed in terms of probabilities. Statistics became a tool of analysis.

The ability to manipulate vast quantities of information in digital form aids calculation and analysis, but it also reveals a fundamental feature of reality. In photography, for example, the detail that could previously only be recorded on a light-sensitive photographic emulsion can now be recorded digitally. The resulting picture 'is' a sequence of binary bits. Similarly, sound can be analysed, stored and transmitted more accurately in digital form than through earlier mechanical or analogue form. A DVD disk can store sound, film, thousands of images or a multivolume encyclopedia. All of these things, which are (in human terms) experienced very differently, are expressed in exactly the same way – by a sequence of binary code, a sequence of off/on gate switches. The fundamental implication of this would be quite astounding were it not so commonplace now, thanks to technology. Reality is built up through a sequence of pieces of information. Everything can be reduced to, and then reconstructed from, the most basic of all forms – a sequence of binary code.

Insight

This chapter has attempted no more than a brief look at some features of the history of science in order to illustrate issues that arise for the philosophy of science. Above all, we need to be aware that, historically, most theories have eventually proved to be wrong, or at least to need adjustment; even the best science cannot yield infallible truths. That, as we turn to questions about how scientific claims may be justified, is an important lesson from history.

KEEP IN MIND...

1 Pre-Socratic thinkers were already asking fundamental questions about the nature of physical reality.

2 Aristotle categorized the sciences, contributed greatly to scientific language and introduced the 'four causes'.

3 Archimedes conducted experiments and produced technology.

4 The medieval world was not just a time of superstition, but produced some good philosophy and science.

5 Francis Bacon, warning of the 'idols' that inhibit knowledge, set criteria for objectivity.

6 With Copernicus and Galileo, observation and simplicity challenged established authority.

7 Newton established laws of physics that were to dominate scientific thinking during the next two centuries.

8 In the nineteenth century, evolution revolutionized our perception of the nature and status of humankind, and statistics started to be employed as a tool of investigation.

9 Twentieth-century science showed the limitations of the Newtonian world-view.

10 The history of science is a useful framework for appreciating what the philosophy of science has to offer.

3

Reasoning and evidence

In this chapter you will:
- *consider the inductive method of argument*
- *examine the place of experiments in science*
- *consider what counts as valid science.*

In this chapter we shall look at a basic approach to science that developed from the seventeenth century and which may still be used to some extent to distinguish between genuine science and pseudoscience. We will be concerned mainly with the inductive method of gaining knowledge, and the impact it had on scientific methodology. The key feature here is the recognition that all claims to scientific knowledge must be supported by evidence from observation and/or experiment.

Subsequent chapters will then consider the debates about how scientific theories are developed, assessed and (if found inadequate) replaced, and the more general problem of scientific realism – in other words, whether scientific statements describe reality or just our experience of reality. A key question in all this is whether scientific theories can ever be proved to be 'correct', in any absolute sense, by the evidence that supports them, and – if not – how we judge between competing theories.

The rise of modern science brought with it an ideal about achieving certainty through reasoning from evidence, but we shall see that, although for some it may remain an ideal, it is very difficult (perhaps impossible) to achieve fully in practice.

Observation and objectivity

Observation has always been the starting point for scientific enquiry. Before any theory is put together, before even the basic questions have been asked, it is the fact that people have observed the natural world and wondered why it is as it is that has been the impetus behind science. Fundamental to the philosophy of science is the discussion of how you move from observations and experimental evidence to theories that attempt to explain what has been observed.

Caution is a keynote in making scientific claims; everything should be backed up by sound theoretical reasoning and experimental evidence. Here, for example, is a statement made by Professor Neil Turok in 1998, in a newspaper article describing his work with Professor Stephen Hawking on the early states of the universe. It contains two very wise notes of caution:

> *First, the discovery is essentially mathematical, and is formulated in the language of the theory of general relativity invented by Albert Einstein to describe gravity, the force which shapes the large-scale structure of the universe. It is hard to describe such things in everyday terms without being misleading in some respects – the origin of our universe was certainly not an everyday event.*
>
> *The second important warning I have to give is that the theories we have built of a very early universe before the Big Bang are not yet backed up by experiment. We often talk as if they are real because we take them very seriously, but we certainly have no special oracular insight to the truth. What we are doing is constructing hypotheses which conform to the very rigorous standard of theoretical physics. But we are under no illusions that, until our theories are thoroughly supported by detailed experimental and observatory results they will remain speculative.*

> *The Daily Telegraph*, 14 March 1998

Notice two important points here:

1 It is not always possible to describe things in language which will enable a non-scientist to get an accurate, imaginative grasp of what is being discussed. Some things are so extraordinary that they make sense only in terms of mathematical formulae.

2 Any hypothesis, however carefully put together based on the best existing theories, must remain provisional until backed up by observations or experimental evidence.

When Galileo argued in favour of the Copernican view of the universe, his work was challenged by the more conservative thinkers of his day, not because his observations or calculations were found to be wrong, but because his argument was based on those observations rather than on a theoretical understanding of the principles that should govern a perfectly ordered world.

The other key difference between the experiments and observations carried out by Galileo and the older Aristotelian view of reality was that Galileo simply looked at what happened, not at why it happened. Observation comes prior to explanation, but it should also take precedence over any assumptions that the observer may have. In other words, the act of observing should be done in as objective fashion as possible.

We have already seen (in Chapter 2) the need for objectivity featured in the work of Francis Bacon, who argued that knowledge should be based on evidence. His 'idols' of habit, prejudice and conformity, and his insistence that one should accept evidence even where it did not conform to one's expectations, mark a clear shift to what became established as the scientific method.

In other words, we should try to set aside all personal preferences and prejudices when observing. But this, of course, is impossible; everything we observe depends on our senses and their particular limitations; everything is seen from a particular perspective. The basic problem with scientific observation is that, as an ideal, it should be absolutely objective, but in reality it will always be limited and prejudiced in some way. Nevertheless, the challenges of the scientific method is to try to eliminate all personal and subjective influences.

Insight

More controversially, as we shall see later, it may involve attempting to set aside the whole web of established knowledge, in order to avoid slotting a new piece of evidence into an existing theoretical mould.

Epistemology is the philosophical term for the study of the theory of knowledge. Some philosophers have argued that all knowledge starts with the operations of the mind, pointing to the ambiguous nature of sense experience; others (empiricists) have argued that everything we know is based on experience.

John Locke (1632–1704) was an empiricist who argued that everything we know comes from sense experience, and that the mind at birth is a blank sheet. Locke divided the qualities we perceive in an object into two categories – primary and secondary:

▶ **primary qualities** – these belonged to the object itself, and included its location, its dimensions and its mass. He considered that these would remain true for the object no matter who perceived it.
▶ **secondary qualities** – these depended upon the sense faculties of the person perceiving the object, and could vary with circumstances. Thus, for example, the ability to perceive colour, smell and sound depends upon our senses; if the light changes, we see things as having a different colour.

Science was therefore concerned with primary qualities. These it could measure, and seemed to be objective, as opposed to the more subjective secondary ones.

Comment

Imagine how different the world would be if examined only in terms of primary qualities. Rather than colours, sounds, tastes, you would have information about dimensions. Music would be a digital sequence, or the pulsing of sound waves in the air. A sunset would be information about wavelengths of light and the composition of the atmosphere.

In general, science deals with primary qualities. The personal encounter with the world, taking in a multitude of experiences simultaneously, mixing personal interpretation and the limitations of sense experience with whatever is encountered as external to the self, is the stuff of the arts, not of science.

Science analyses, eliminates the irrelevant and the personal, and finds the relationship between the primary qualities of objects.

Setting aside the dominance of secondary qualities in experience, along with any sense of purpose or goal, was essential for the development of scientific method – but it was not an easy step to take. The mechanical world of Newtonian physics was a rather abstract and dull place – far removed from the confusing richness of personal experience.

The more we look at the way information is gathered and described, the clearer it becomes that there will always be a gap between reality and description. Just as the world changes depending upon whether we are mainly concerned with primary or secondary qualities, so the pictures and models we use to describe it cannot really be said to be 'true', simply because we have no way to make a direct comparison between the description and model we use and the reality to which it points. Our experience can never be unambiguous, but the methods developed by science aimed to eliminate personal factors as much as possible.

Are instruments objective?

Even instruments can cause problems. For example, using his telescope, Galileo saw that there were mountains on the Moon, a discovery that challenged the received tradition that the heavenly bodies were perfect spheres. However, this is not simply a triumph of evidence over metaphysical prejudice, since – from the drawings Galileo made – we know that some of his observations were wrong. Some of what he perceives as mountains must have been the result of distortions in the glass of his telescope.

Hence, the observations or experimental data used by science are only going to be as good as the equipment and instruments used to gather them.

As we shall see, the recognition that we cannot simply observe and describe came to the fore in the twentieth century, particularly in terms of subatomic physics, since it seemed impossible to disentangle what was seen from the action of seeing it.

Induction

The rise of science was characterized by a new seriousness with which evidence was gathered and examined in order to frame general theories. There are two very different forms of argument:

▶ A **deductive** argument starts with a general principle and deduces other things from it. So, for example, if you assume that all heavenly bodies must be perfect spheres, a logical deduction is that there cannot be mountains on the Moon.
▶ An **inductive** argument starts with observations or experimental results and on the basis of these sets about framing general principles that take them into account. Thus, observing mountains on the Moon, you conclude that not all heavenly bodies are perfect spheres.

It was the inductive form of argument – generally termed 'inductive inference' – that became a distinguishing feature of modern science.

Bertrand Russell described the 'principle of induction' by saying that the more two things are observed together, the more it is assumed that they are causally linked. If I perform an experiment only once, I may be uncertain of its results. If I perform it a hundred times, with the same result each time, I become convinced that I will obtain that result every time I perform it. Thus far, this sounds no more than common sense, but it raises many problems, for it is one thing to anticipate the likely outcome of an experiment on the basis of past experience, quite another to say that the past experience proves that a certain result will always be obtained, as we shall see later.

The black swan

There is a classic example of induction which makes the situation so clear that it is always worth repeating...

Someone from Europe, having seen many swans, all of them white, comes to the conclusion that 'all swans are white' and anticipates that the next swan to appear will also be white. That generalization is confirmed with every new swan that is seen. Then, visiting Australia, the person comes across a black swan and has to think again.

The generalized conclusion from seeing the white swans is as example of what is termed 'enumerative induction' – as the numbers stack up, so the conclusion seems more and more certain. But the key thing to recognize is that, while the evidence suggests that all swans are white, it cannot prove that all swans are white, as the first encounter with a black swan shows. Of course, this example is clear because 1) swans are easily distinguished from other birds and 2) adult swans come in suitably contrasting colours. In most other situations we may find that there is doubt about the classification of the thing we are examining (improbably: 'Is it a swan or could it be a duck or a goose?') and the claim to see an anomaly might not be so black or white!

THE INDUCTIVE METHOD

The inductive approach to knowledge is based on the impartial gathering of evidence, or the setting up of appropriate experiments, such that the resulting information can be examined and conclusions drawn from it. It assumes that the person examining it will come with an open mind, and that theories framed as a result of that examination will then be checked against new evidence.

This is sometimes presented as the general 'scientific method', although we need to remember that science works in a number of different ways. In practice, induction works like this:

▶ **Evidence** is gathered, and irrelevant factors are eliminated as far as possible.
▶ Conclusions are drawn from that evidence, which lead to the framing of an **hypothesis**.
▶ **Experiments** are devised to test out the hypothesis, by seeing if it can correctly predict the results of those experiments.
▶ If necessary, the hypothesis is modified to take into account the results of those later experiments.
▶ A **general theory** is framed from the hypothesis and its related experimental data.
▶ That theory is then used to make **predictions**, on the basis of which it can be either confirmed or disproved.

Example

The final step in this process is well illustrated by the key prediction that confirmed Einstein's theory of General Relativity. Einstein argued that light would bend within a strong gravitational field, and therefore that stars would appear to shift their relative positions when the light from them passed close to the Sun. This was a remarkably bold prediction to make. It could be tested only by observing the stars very close to the edge of the Sun as it passed across the sky, and by comparing this with their position relative to other stars once the light coming from them was no longer affected by the Sun's gravitational pull. But the only time when they could be observed so close to the sun was during an eclipse. Teams of observers went to Africa and South America to observe an eclipse in 1919. The stars did indeed appear to shift their positions to a degree very close to Einstein's predictions, thus confirming the theory of General Relativity.

Confirmed scientific theories are often referred to as 'laws of nature' or 'laws of physics', but it is important to recognize exactly what is meant by 'law' in this case. This is not the sort of law that has to be obeyed. A scientific law cannot dictate how things should be; it simply describes them. The law of gravity does not require that, having tripped up, I should adopt a prone position on the pavement – it simply describes the phenomenon that, having tripped, I fall. Hence, if I trip and float upwards, I am not disobeying a law, it simply means that I am in an environment (e.g. in orbit) in which the phenomenon described by the 'law of gravity' does not apply, or that the effect of gravity is countered by other forces than enable me to float upwards. The 'law' cannot be 'broken' in these circumstances, only be found to be inadequate to give a complete description of what is happening.

A CLASSIC CRITIQUE OF EMPIRICAL EVIDENCE

The philosopher David Hume pointed out that scientific laws were only summaries of what had been experienced so far. The more evidence that confirmed them, the greater their degrees of probability, but no amount of evidence could lead to the claim of absolute certainty.

Hume argued that the wise man should always proportion his belief to the evidence available; the more evidence in favour of something (or balanced in favour, where there are examples to the contrary), the more likely it is to be true. He also pointed out that, in assessing evidence, one should take into account the reliability of witnesses, and whether they had a particular interest in the evidence they give. Like Francis Bacon, Hume sets out basic rules for the assessment of evidence, with the attempt to remove all subjective factors or partiality, and to achieve as objective a review of evidence as is possible.

What Hume established (in his *Enquiry Concerning Human Understanding*, section 4) was that no amount of evidence could, through the logic of induction, ever establish the absolute validity of a claim. There is always scope for a counter-example, and therefore for the claim to fail.

Insight

This gets to the heart of the 'problem of induction' and raises the most profound problems for science since it challenges the foundations of the scientific method. Many of the later discussions about how to deal with competing theories, or the limitations of scientific claims, stem from this basic problem – evidence cannot yield absolute certainty.

With hindsight, that might seem a very reasonable conclusion to draw from the process of gathering scientific evidence, but in Hume's day – when scientific method was sought as something of a replacement for Aristotle in terms of a certainty in life – it was radical. It was this apparent attack on the rational justification of scientific theories that later 'awoke' the philosopher Kant from his slumbers. He accepted the force of Hume's challenge, but could not bring himself to deny the towering achievements of Newton's physics, which appeared to spring from the certainty of established laws of nature. It drove Kant to the conclusion that the certainty we see in the structures of nature (particularly in the ideas of time, space and causality) are there because our minds impose such categories upon the phenomena of our experience.

In many ways, the distinction Kant made between the 'phenomena' of our experience and the 'noumena' of things as they are in themselves, with the latter unknowable directly, continues to be relevant. I cannot know an electron as it is in itself, but only as it appears to me through the various models or images by which I try to understand things at the subatomic level. I may understand something in a way that is useful to me, but that does not mean that my understanding is – or can ever be – definitive.

John Stuart Mill (1806–73), a philosopher best known for his work on freedom and on utilitarian ethics, gave an account of how we go about using inductive reasoning in his book *A System of Logic, Ratiocinative and Inductive* (1843), using the principles of similarity and difference. Broadly, the argument works along these lines:

Suppose I want to examine why it is that people contract a particular illness...

I look at all the evidence about that incidence of that illness to see if there are any common antecedent factors that might have been its cause. Do they all eat a particular diet? Do they smoke? Do they come from a particular part of the world? Do they all use the same source of drinking water? In other words, I am looking for similarities in the background of the cases I am examining.

But not all of those similarities will be relevant. The fact that they are all human beings need not be taken into account, because that's equally shared by those who do not have the illness. But perhaps one might look at their racial group or their age. If a factor is not common to the group as a whole, it is unlikely to be relevant.

Equally, I can examine differences. Given a number of people, only one of whom has the illness I am examining, I can look for differences in their background. If I find something about the person with the illness that does not apply to all of those who have not contracted it, that becomes a likely cause. If all the circumstances except one are held in common, then that one is the cause of the illness.

Insight

The problem, of course, is that you seldom have a situation where you can be sure that there is only one difference in circumstances, and therefore that you have found the cause of what you are examining. The result may be brought about by a factor you have not yet thought of checking, or by a particular arrangement of otherwise common circumstances.

It will be clear by now that this kind of reasoning can never give certainty, but only an increasingly high degree of probability. There is always going to be the chance that some new evidence will show that the original hypothesis, upon which a theory is based, was wrong. Most likely, it is shown that the theory only applies within a limited field, and that in some unusual set of circumstances it breaks down. Even if it is never disproved, or shown to be limited in this way, a scientific theory that has been developed using this inductive method, is always going to be open to the possibility of being proved wrong. Without that possibility, it is not scientific. This is the classic problem of induction – no amount of evidence will ever be enough to prove the case.

In an example in his *Problems of Philosophy* (1952), Bertrand Russell gives a characteristically clear and entertaining account of the problem of induction. Having explained that we tend to assume that what has always been experienced in the past will continue to be the case in the future, he introduces the example of the chicken which, having been fed regularly every morning, anticipates that this will continue to happen in the future. But, of course, this need not be so:

The man who has fed the chicken every day throughout its life at last wrings its neck instead, showing that more refined views as to the uniformity of nature would have been useful to the chicken.

Bertrand Russell, *Problems of Philosophy* (1952)

An important modern discussion of the problem of induction was set out by Professor Nelson Goodman of Harvard in 1955, in his influential book *Fact, Fiction and Forecast*, and the examples he gave (e.g. 'All ravens are black' and the colour 'grue') are frequently cited in other books.

Goodman takes Hume's view that there are no necessary connections between matters of fact. Rather, experiencing one thing following another in a regular pattern leads us to a habit of mind in which we see them associated and therefore to claim that one causes the other. Everything is predicted on the basis of past regularity, because regularity has established a habit.

We establish general rules on the basis of particulars that we experience, and those rules are then used as the basis for inference – in other words, observation of particular events lead to a rule, and the rule then leads to predictions about other events. The important thing is to realize that the past can place no logical restrictions on the future. The fact that something has not happened in the past does not mean that it cannot happen in the future.

Notice the circularity in the way induction is used – rules depend on particulars and the prediction of particulars depends on rules. We justify the 'rules of induction' by saying that they are framed on the basis of successful induction. That's fine for practical purposes, but it does not give any independent justification for predictions about future events. It works because it works; but that does not mean that it *has* to work. Goodman comments:

> **A rule is amended if it yields an inference we are unwilling to accept; an inference is rejected if it violates a rule we are unwilling to reject.**

Nelson Goodman, *Fact Fiction and Forecast* (4th edn), page 64

The only justification therefore lies in the apparent agreement between rules and inferences: if a rule yields acceptable inferences, it is strengthened. The crucial question, according to Goodman, is not how you can justify a prediction, but how you can tell the difference between valid and invalid predictions.

He makes the important distinction between 'law-like' statements and accidental statements. If I use a particular example in order to support an hypothesis, that hypothesis must take the form of a general law (whether it is right or not is another matter). To use his examples:

▶ I can argue from the fact that one piece of copper conducts electricity to the general principle that all copper conducts electricity.
▶ But I cannot argue from the fact that one man in a room is a third son to the hypothesis that every man in the room is a third son. Being a third son in this situation is just something that happens to be the case in this instance – it is not a general feature of humankind, in the way that conducting electricity is a general feature of copper.

Here is the famous problem of 'grue':

▶ All emeralds examined before time 't' are green – therefore you reach the conclusion that all emeralds are green.
▶ But suppose you use the term 'grue' for all things examined up to time 't' that are green, and all other things that are blue.
▶ In this case, up to time 't', all emeralds are both green and grue; after 't' an emerald could only be grue if it was in fact blue.
▶ Now the problem is that, up to time 't', our usual approach to induction confirms 'all emeralds are green' and 'all emeralds are grue' equally – and yet we know that (after time 't') the first is going to be true and the second false. How, up to that point, can we decide between them?

In other words

From the standpoint of the inductive method, there is, prior to time 't', no way of deciding between emeralds being green and emeralds being 'grue' – both, on the evidence, are equally likely. But we know, of course, that one is very soon going to be wrong and the other right. Hence, there is a major weakness in the use of induction in order to predict what will be the case in the future.

Now the key feature here is that an 'accidental hypothesis' (unlike a law-like hypothesis) has some restriction in terms of time or space.

In other words, it cannot be generalized. The problem with 'grue' is that it has a temporal restriction, in that it means one thing before a particular time and something else after it. The new riddle of induction is not so much Hume's problem about how you justify general laws in terms of individual cases, but how you tell those hypotheses that can correctly be generalized from particular instances and those which cannot.

Let us consider one final example to illustrate the problem of induction:

▶ Can induction prove the statement 'All planets with water flowing on their surface are likely to support life'?
▶ We know, in the case of Earth, that it is correct. But is that a general feature of planets of a certain size and distance from their suns, or is it simply an accidental feature of our own planet?

The big issue is that science looks for general features and principles, which have to be abstracted out of the particulars in which we encounter them. We have encountered life on only one planet – our own. But we cannot yet know whether that is an accident, and therefore possibly unique, or whether it is a general feature of planets of certain types. If we discover water in liquid form on another planet, but do not discover life there, the statement is disproved; if we do also find life, the statement is strengthened, but only to the extent that there are two actual examples, not that all such planets support life.

A mathematical universe

It is one thing to observe nature, another to explain it, and one of the key components in the explanations given by scientist in the seventeenth and eighteenth centuries was mathematics. Galileo thought that the book of nature was written in the language of mathematics, but this was not a new idea, for Pythagoras (570–497 BCE) had argued that everything could be given an explanation in terms of mathematics. Even the title of Newton's most famous book is *Philosophiae naturalis principia mathematica* – an attempt to understand the workings of nature on mathematical principles. Work in mathematics thus provided the background to much of the advancement of science in the seventeenth and eighteenth centuries.

It is important to recognize the nature of mathematics and the very radical abstraction that it involves. Galileo, Descartes, Huygens and Newton all produced formulae. In other words, they were seeking to create a mathematical and abstract way of summing up physical phenomena, using mathematics to express patterns seen in nature. That it should be possible for an abstract formula to correspond to nature was a fundamental assumption made by those involved in the emerging sciences. Beneath it lay the deeper assumption that the world is a predictable and ordered place. Escaping from the earlier era of crude superstition and magic, they saw themselves emerging into a world where reason and evidence would triumph. But reason, in its purest form, is seen in logic and mathematics, and it was therefore natural to expect that the world would be, in principle, comprehensible in terms of 'laws of nature' which, with mathematical precision, would determine the movement of all things.

The result of this was that the science produced in this period was not about what is experienced – with all its mixtures of sensations, beauty, sounds, etc. – but the abstract formulae by which such things could be understood and predicted. Phenomena were thus 'reduced' to their mathematically quantifiable components.

We have already explored this briefly in looking at John Locke's distinction between primary and secondary qualities. Fully aware that colour, sound and taste were obviously linked to the human sense organs, he called them 'secondary' qualities. The primary ones were mass, location and number – exactly those things that can be measured and considered mathematically. By the end of the seventeenth century science thought of 'real' nature as silent, colourless and scentless, an interlocking network of material bodies, whose activities could be quantified, analysed and expressed in the form of scientific and mathematical laws.

Notice how abstract the very concept of number is. I see three separate objects before me and describe them as being 'three'. Yet there is nothing in the description of each of them that has the inherent quality of 'threeness'. 'Three' is a purely abstract term, used in order to sum up a particularly useful feature of that experience. Thus, if I am receiving money, it is the number on the banknote that is of prime importance; its colour or the quality of its paper is of less significance. On the other hand, in a collection of green objects, a dollar bill might be quite in place, its numerical value of little significance.

The key thing to remember here is that mathematics is an abstraction, not a reality. A key feature of seventeenth-century science was that the whole scheme of highly abstract reasoning was mistaken for reality itself. Hence it was given an 'objectivity' that led to the assumption that, once all 'laws' had been formulated, there would be nothing left to discover. With the twentieth century and the recognition of the validity of different and even conflicting theories, the attempt to 'objectify' this abstraction process was recognized to be limited.

Insight

Once you identify 'reality' with mathematical formulae or the theories that science abstracts from experience or experimental data, then the world appears to be a mathematically controlled and determined machine. But reality is not science; reality is what science seeks to explain, and our explanations will always be limited, open to be challenged by new ideas and theories.

Experiments

At several points so far we have recognized that scientific evidence comes from experiments as well as from observations. In particular, once a theory has been formulated, it is important to set about finding experiments that will either confirm or refute it.

There are two fundamentally important features of scientific experiments:

1 THE ISOLATION OF SIGNIFICANT VARIABLES

Experiments create an artificial situation in which, as far as possible, all extraneous influences are eliminated, so that the investigator can focus on and measure a single or small number of variables. The more delicate the thing that the experiment is to measure, the more stringent are the safeguards to eliminate external factors. Thus, for example, the experiment to test the presence of the most elusive neutrinos passing through the Earth was conducted using a tank of absolutely pure water buried deep below the surface of the Earth, far from all possible sources of interference.

To illustrate the importance of isolating significant variables, let us take as an example the experimental testing of a new drug. Suppose only those patients who are most seriously ill are given the new drug, and those with a milder condition are given more conventional treatment. The results might well show that, statistically, more people die after taking the new drug. This would not be a valid experiment however, because there is the obvious intrusion of an unwanted variable – namely the severity of the illness. In order for the experiment to be accurate, it would be necessary to make sure that two groups of patients were identified, each having the same mix in terms of age, sex and severity of illness. One group could then be given the new drug and the other would receive either no drug at all, or some more conventional treatment.

The result of that experiment might be to say that the new drug produced X per cent increase in life expectancy. In other words, all other things being equal, this is the benefit of the drug. If it is subsequently found that there were all sorts of other factors of which those conducting the experiment were unaware, then the value of the experiment would be severely reduced.

••

Insight

No experiment can ever show the whole situation; if it did, it would have to be as large and complex as the universe. Experimental evidence is therefore highly selective, and may reflect the assumptions of the scientist. This, as we shall see later in this book, is the root cause of much of the debate about the status of scientific theories.

••

2 THE ABILITY TO REPRODUCE RESULTS

If something is observed just once, it could be a freak occurrence, caused by an unusual combination of circumstances. It would certainly not be an adequate basis on which to frame a scientific hypothesis. The importance of carefully defined experiments is that they enable other people to reproduce them, and thus confirm or challenge the original findings. Once the result of an experiment is published, scientists in the same area of research all over the world attempt to reproduce it in order to see if they get the same results, or to check whether all extraneous variables have in fact been eliminated. If the results cannot be reproduced, they are regarded as highly suspect.

Hence, the devising of suitable experiments plays a hugely important part in the overall activity of science. Planning and organizing an experiment, creating the right conditions and devising and refining measuring equipment, checking that all other variables have been eliminated – these very often constitute the bulk of the work done in many areas of science, compared with which the actual running of the experiment itself may be relatively easy.

Can there be crucial experiments which are decisive in saying which of a number of competing theories is correct? Early scientists (e.g. Francis Bacon) thought this possible, but others (e.g. Pierre Duhem, a physicist writing at the end of the nineteenth century and the first years of the twentieth) have argued that they are impossible, since you can never know the sum total of possible theories that can be applied to any set of experimental results. Hence, even if the experiment is run perfectly, that does not guarantee that there is one and only one possible theory to be endorsed by it. For practical purposes, however, some experiments (e.g. the Eddington observations that confirmed Einstein's theory of General Relativity – see Chapter 3) do appear to be decisive in saying that, of existing theories, one is superior to the others.

Hypothesis and confirmation

So far we have looked at the way a scientist can argue from individual bits of information or experimental results towards a generalized statement, but that is not the only way in which theories are developed and tested against evidence. The **hypothetico-deductive method** operates in this way:

- ▶ Start with an hypothesis
- ▶ Deduce from it a prediction
- ▶ Check whether that prediction is correct
- ▶ If it is correct, the hypothesis is confirmed.

This is the opposite way round from induction, since it starts with a general idea and then devises means of getting information that can confirm that the idea is correct. The important thing, however, is that it should not impose an hypothesis upon the evidence, rather it uses evidence to test out the hypothesis.

However, when it comes to testing out an hypothesis against evidence, there is a problem with knowing what can count for its confirmation. Carl Hempel (1905–97) gave an example that has troubled many other philosophers, it is known as the **'raven paradox'**:

All ravens are black.
This statement predicts that, if you find a raven, it will be black, so every black raven you find will confirm it. So far so good.

However, 'All ravens are black' is the logical equivalent of 'All non-black things are not ravens.' In other words, if the first of these statements is true, the second must also be true – if it's not black, then obviously it's not a raven.

This produced a problem. Suppose I say 'All bananas are yellow' (ignoring, for this argument, that they will eventually go brown). This confirms the statement that all non-black things are not ravens, since the banana is not black and it is not a raven. But, logically, if it confirms that second statement, then it must also confirm the first, since they are logically equivalent. Hence we arrive at the crazy idea that statements about bananas being yellow, or leaves green, or shoes brown, will confirm our original claim that all ravens are black.

One way of trying to get round this problem is to say that all those other statements do indeed confirm the original claim that all ravens are black but that they do so only very weakly. Imagine you go looking for something. As you hunt, you keep discarding everything that is not what you are looking for, so that everything you discard brings you – in theory at least – one step nearer finding that searched-for thing. The yellow banana does confirm the black raven, but only in so far as that yellow thing is not, after all, a raven!

TESTING AGAINST EXPERIENCE

Induction, as a method, may have been fine for some problems, but in dealing with relativity and in assessing quantum theory, Einstein recognized that there was no way of arguing from experience to theory. His method was to start with logically simple general principles, develop a theory on the basis of them, and then test that theory out against experience.

Einstein's wonderful early thought experiments – for example, a flashbulb going off in a railway carriage, illuminating clocks simultaneously (from the perspective of someone in the carriage) or not (from a stationary observer outside the carriage) – present puzzles of a practical sort the answers to which illuminate his general theory of relativity.

Given the theory, the implications for observing clocks in railway carriages gives practical confirmation; but no amount of observing passing trains is ever (on a basis of inductive inference) going to generate the theory of relativity.

Abductive reasoning

We have looked at induction (where evidence is used to support a theory) and deduction (where we examine the logical consequence of a theory), but very often what we are trying to do is find the best possible explanation for something, and use a form of reason that is called **abduction**. Consider this example:

I see an unopened letter on the pavement early in the morning and reckon that the best explanation is that the postman dropped it.

I infer, from the letter on the pavement, that the postman dropped it because, if the postman dropped it, it would be there on the pavement. In other words, I am working from a consequence back to a precondition.

Of course, it is possible that someone else dropped it, but of the infinite number of possible reasons for the letter being there, I have attempted to narrow down the possibilities in order to test out and see if the one I have inferred is correct. Science often works by seeking the 'inference to the best explanation', using this kind of abductive reasoning.

What counts as science?

Blind commitment to a theory is not an intellectual virtue: it is an intellectual crime.

Imre Lakatos, 1973

Science always requires a healthy measure of scepticism, a willingness to re-examine views in the light of new evidence, and to strive for theories that are based on objective facts that can be checked and confirmed. As we saw earlier, it was the quest for objectivity, loyalty to the experimental method, and a willingness to set aside established ideas in favour of reason and evidence, that characterized the work of Francis Bacon and others. There were disagreements about the extent to which certainty was possible, and some (e.g. Newton) were willing to accept 'practical certainty' even though recognizing that 'absolute certainty' was never going to be possible.

The assumption here is that, if you want to explain a phenomenon, you need to show the conditions that have brought it about, along with a general 'law of nature' that it appears to be following. This is sometimes referred to as the **'Covering Law' model**. That would be fine, except that the general law is framed on the basis of evidence that is never going to be complete, and the phenomenon you are attempting to explain may well be a bit of evidence that will eventually show your 'law' to be inadequate.

In the twentieth century there was considerable debate, as we shall see in Chapter 4, about the process by which inadequacies in a theory are examined, and the point at which the theory should be discarded. No scientific theory can be expected to last for all time. Theories may be falsified by new and contrary evidence (Popper) or be displaced when there is a general shift in the matrix of views in which they are embedded (Kuhn).

On this basis, we cannot say that good or effective science provides knowledge that will be confirmed as true for all time, whereas bad or ineffective science yields results that have been (or will be) proved false. After all, something that is believed for all the wrong reasons may eventually be proved correct, and the most cherished theories in science can be displaced by others that are more effective. What distinguished science from other methods of enquiry is to do with the nature of the claims it makes, and the methods by which it seeks to establish them.

KEEP IN MIND...

1 Reasoning from evidence was a key feature of the rise of modern science.

2 We cannot be absolutely certain about any theory that is based on evidence, since new evidence may always appear to undermine it.

3 The inductive method argues from evidence to theory, not vice versa.

4 Hume argued that belief should always be proportional to evidence.

5 Goodman pointed to the problem of 'grue' where evidence confirms two contradictory claims.

6 Newton saw the world as operating on rational and mathematical principles.

7 Experiments seek to provide measurable data with a limited number of variables.

8 An abductive argument infers a reason in order to explain a present phenomenon.

9 Science is defined by its methodology rather than its content.

10 The willingness to allow evidence to confirm or challenge theories is a mark of genuine science.

4

Scientific theories and progress

In this chapter you will:
- *consider the status of scientific claims*
- *consider the way in which scientific theories may be challenged*
- *examine the way science makes progress.*

In Chapter 3 we examined the inductive method of reasoning that was basic to the establishment of modern science. We saw, however, that it presented various problems, so that the process by which theories were established and revised was far from straightforward. During the twentieth century the place of the inductive method in science was challenged and there was considerable debate about how scientific theories should be evaluated, how science makes progress, how one law or set of laws comes to replace another, and whether (if at all) one can ever decide that one particular theory is inherently better than another.

In this chapter we shall look at various approaches to this set of problems. Of particular importance here is the work of Karl Popper and Thomas Kuhn, but we shall also look briefly at Paul Feyerabend and Imre Lakatos. Two closely related questions have chapters of their own: the issues of scientific realism and of relativism and objectivity. In examining them, we shall need to cover similar ground, but with those specific questions in mind.

Background

Until the end of the nineteenth century the general scientific view of the world was that it was a mechanism in motion, waiting for humanity to measure and calculate its operation. It was thought of as being quite independent of the person perceiving it, with laws of nature that determined its operations and the interaction of all its parts. Science assumed that it had set aside all metaphysical speculation and would eventually achieve a full and systematic knowledge of the physical world. There were to be no a priori elements in scientific knowledge – all presuppositions or hypotheses could be set aside in favour of theories based on solid evidence and proved by inductive argument. The scientist was required to take an objective look at matter, and formulate theories to explain its operations. The nineteenth-century philosopher Ludwig Büchner (1824–99) could therefore make the general assertion that 'There is no force without matter; no matter without force' without fear of contradiction, and Ernst Haeckel was confident that science had already revealed almost all there was to know about the world and had displaced the crude superstitions of earlier days.

But alongside this very confident view, another was developing that eventually came to dominate scientific thinking in the early twentieth century. Immanuel Kant, the eighteenth-century German idealist philosopher, had argued for distinction to be made between things in themselves (**noumena**) and things as we perceive them to be (**phenomena**). All the evidence we receive from our senses is 'phenomena'; we only know what we perceive. We may assume that there is a separate reality 'out there' causing us to have those sensations, but – if so – we can never engage it directly.

Thus, from the work of Hermann von Helmholtz (1821–94) in the 1870s through to Ernst Cassirer (1874–1945) in the early years of the twentieth century, there developed a recognition that science is not looking at things in themselves, but at the structures of phenomena – the way in which we perceive the world.

The philosopher and logician Ernst Mach (1838–1916), in *The Analysis of Sensations* (1886), argued that science reflects the content of the consciousness, as it is produced by sensation. There are no predetermined structures, but everything should be reducible to statements about sensations. The only exception to this was his acceptance of logical and mathematical propositions. So, from his perspective, a scientific theory was the description of some regularity within the phenomena of sensations.

THE LOGICAL POSITIVISTS

Into this situation there came the Logical Positivists of the Vienna Circle, of whom probably the best known are Moritz Schlick (1882–1936) and Rudolf Carnap (1892–1970). They were generally interested in both science and mathematics and were influenced by the work of the early Ludwig Wittgenstein (1889–1951) and Bertrand Russell. Their task was to determine what constituted a valid, factual proposition. They wanted to define correspondence rules, by which the words we use relate to observations. They also wanted to reduce general and theoretical terms (e.g. mass or force) to those things that could be perceived. In other words, they thought that the mass of a body should be defined in terms of the measurements that can be made of it.

In general, the position adopted by the Logical Positivists was that the meaning of a statement was its method of verification. If I say that there is a dog in the room, I actually mean that, if you go into the room, you will see a dog. If you cannot say what evidence would count for or against a proposition – in other words, how you could verify it through sense experience – then that proposition is meaningless.

Now, clearly, this is mainly concerned with the use of language, and on that basis they were happy to reject the traditional claims of both metaphysics and ethics. But for science it had a particular importance, for it assumed that the only valid statements were those confirmed by experimental or observational evidence.

Insight

The Logical Positivists saw science as tracking patterns in experience, concerned only with things that could be observed. In particular, they rejected any idea of deeper structures or levels of meaning – everything was on the surface, awaiting literal description.

This seems a logical development of the traditional inductive method of science, but it raises some fundamental questions:

▶ What do you do if faced with two alternative theories to explain a phenomenon? Can they both be right?
▶ Can everything science wants to say be reduced to sensations?

Once a theory is accepted and established as a 'law of nature' on the basis of empirical evidence, it seemed inconceivable that it could subsequently be proved wrong. Theories which apply to a limited range of phenomena might be enlarged in order to take into account some wider set of conditions, but it was difficult – given the strict process of induction from the phenomena of sense experience – to see why a theory, so confirmed by experience, should be set aside in favour of another.

Insight

However, even while this view was dominating the philosophy of science, the actual practice of science – especially in the fields of relativity and quantum physics – was producing ideas that did not fit this narrow schema.

Some of the important thinking about scientific theories and how they develop and get replaced was essentially a reaction against this accepted view of the place of induction from sense experience. In this chapter we shall now turn to the work of Popper, who criticized Logical Positivism and showed the role of falsification in the examination and replacement of theories. But there was also a sense (as exemplified in the work of Kuhn) that scientific theories were framed within an overall view of the world (a *Weltanschauung*), and that radical change could take place only when one whole set of views was finally found to be inadequate and replaced by another.

Falsification

Karl Popper (1902–94) was an Austrian philosopher from Vienna who, having spent some years in New Zealand, settled in London in 1945, where he became Professor of Logic and Scientific Method at the London School of Economics. He made significant contributions to political philosophy as well as to the philosophy of science.

Karl Popper's **theory of falsification**, although important for the philosophy of science, has much wider application. In the 1920s and 30s Logical Positivists were arguing that statements only had meaning if they could be verified by sense data. In other words, if you could not give any evidence for a statement, or say what would count for or against it, then it was meaningless. (The exception, of course, was statements of logic or mathematics where the meaning is already contained within the definition of the words used. You don't have to go out and point to things in order to show that $2 + 2 = 4$.)

In his book *The Logic of Scientific Discovery* (1934; translated in 1959), Popper argued that one could not prove a scientific theory to be true simply by adding new confirming evidence. On the other hand, if some piece of sound evidence goes against a theory, that may be enough to show that the theory is false.

He therefore pointed out that a scientific theory could not be compatible with all possible evidence. If it is to be scientific, then it must be possible, in theory, for it to be falsified. In practice, of course, a theory is not automatically discarded as soon as one piece of contrary evidence is produced, because it might be equally possible that the evidence is at fault. As with all experimental evidence, a scientist tries to reproduce this contrary evidence, to show that it was not a freak result but a genuine indication of something for which the original theory cannot account.

Insight

In other words, rather than simply adding up bits of evidence, Popper argued that science makes progresses by putting forward bold conjectures, followed by a process of radical testing to see if they can be falsified.

Popper was particularly critical of the Marxist theory of dialectical materialism and Freudian psychology. He observed that Marxists have the habit of interpreting every event in terms of Marxist theory, and then using such interpretations to produce more evidence to confirm that theory. He argued that, if nothing were allowed to falsify the Marxist view of dialectical materialism, then that theory could not be genuinely scientific. Similarly, he suggested that a psychologist might be tempted to give a particular interpretation of a patient's condition, based on the accepted theory, and to attempt to explain away or ignore anything which does not appear to fit the theoretical expectations.

A key feature of Popper's claim here is that scientific laws always go beyond existing experimental data and experience. The inductive method attempted to show that, by building up a body of data, inferences can be made to give laws that are regarded as certain, rather than probable. Popper challenges this on the grounds that all sensation involves interpretation of some sort, and that in any series of experiments there will be variations, and whether or not such variations are taken into account is down to the presuppositions of the person conducting them. Also, of course, the number of experiments done is always finite, whereas the number of experiments not yet done is infinite, so an inductive argument can never achieve the absolute certainty of a piece of deductive logic. At the same time, scientists are likely to favour any alternative theories that can account for both the original, confirming evidence and also the new, conflicting evidence. In other words, progress comes by way of finding the limitations of existing scientific theories and pushing beyond them.

For Popper, all genuine scientific theories had to be logically self-consistent and capable of falsification; an irrefutable theory is not scientific.

Example

Until Einstein's prediction that light from distant stars would bend due to the Sun's gravitational pull was confirmed by observations during an eclipse, it was believed that light always travelled in straight lines. The original view was therefore modified to take the effect of strong gravitational forces into account. Hence the earlier theory, being genuinely scientific, is capable of being shown to apply only within a limited range of conditions.

Popper's view challenged two popular philosophical ideas:

1 Locke's idea that the mind is a tabula rasa until it receives experience.
2 Wittgenstein's idea, in the *Tractatus Logico-Philosophicus* (1921), that the task of language is to provide an image of the external world.

Instead, he saw the human mind as having a creative role vis-à-vis experience. In the scientific realm this means that progress is made when a person makes a creative leap to put forward an hypothesis that goes beyond what can be known through experience. It does not progress gradually by the adding up of additional information to confirm what is already known, but by moving speculatively into the unknown, and testing out hypotheses, probing their weak points and modifying them accordingly.

This view of scientific work paralleled Popper's more general view that the basic form of intellectual work is problem-solving.

Insight

The need for critical engagement is central for Popper's politics as well as his philosophy of science. Democracy allows people to engage with, criticize and replace governments, whereas authoritarian systems impose answers and forbid questions.

In effect, the goal of science is therefore to produce propositions which are high in information content, and which are therefore low in probability of being true (since the more information they contain, the greater the chance of finding them to be false), but which actually come close to the truth. It would, of course, be easy to find a statement that never need fear being refuted (e.g. 'The sun will rise tomorrow'), but it offers so little informational content that it is difficult to see how it can be of much practical use.

Popper's approach to scientific method was therefore as follows:

1 be aware of the **problem** (e.g. the failure of an earlier theory)
2 propose a **solution** (i.e. a new theory)
3 deduce **testable propositions** from that theory
4 establish a **preference** from among competing theories.

Therefore, in terms of the results of scientific work, he observes that everything is already 'theory soaked' and is a matter of using and constantly modifying theories in the light of available evidence.

In general, science works by means of experiments. Results are presented along with detailed information about the experimental methods by which they have been obtained. The task of those who wish to examine the results is to repeat the experiments and see if they produce identical results. Now, as it goes on, a theory is going to predict facts, some of which will be verified, some of which will not. Where it has failed to predict correctly, it may need to be replaced. However, it is unlikely that a theory will be discarded at the first hint of falsification. On the other hand, when an alternative theory becomes available, every occasion of falsification leads to a comparison between the two theories, and the one that is confirmed more broadly is the one to be accepted.

Certainty or quest?

Is there a solid basis upon which all our knowledge can be shown to rest? The quest for such a basis is generally termed **foundationalism.** Some seek it in pure reason, others in sense experience; others opt for a foundation in terms of religious or other authority. Perhaps, if empirical evidence cannot yield certainty (for reasons that Hume and others have shown), then science may need to be its own judge of truth – by opting for one overall paradigm rather than another (see below), or seeing truth as related only to a particular research project and evaluated within its own terms.

If every theory needs to be open to the possibility of being falsified, does that mean that there is no 'truth' or 'reality', but it is all down to whatever particular theory we happen to accept at this moment in time? Is it possible to do science without some certainty upon which to base one's work?

Popper's answer to this is both radical and remarkable. He argued for 'rationality without foundations' (to quote the subtitle of a very useful book on Popper by Stefano Gattei; see the Further reading at the end of this book). He points out that we can opt to take rational argument as the basis for our quest for truth without having any proof or guarantee that rational argument is a solid foundation. Given a scientific theory or a political regime, we can opt to examine it critically, modify it if necessary, but yet remain open to accept criticism of our own reasoning.

In taking that view, we can believe that there is 'truth' or 'reality' as a regulative idea – as something that guides our quest, but about which we can never claim to have reached absolute truth or reality, recognizing that our attempts to describe it will always be fallible and open to falsification. What Popper advocates is an attitude and a commitment, undertaken without any guarantee and with a recognition of our limitations.

Hence, he would see the scientist as someone who is always trying to find the truth about the reality of what he or she is examining, but who recognizes the limitations of what is known so far, is positively critical about his or her views as well as those of others, and who hopes that the process of critical engagement with the problem will move knowledge further along in the direction of a reality which will nevertheless encompass more than we are ever capable of knowing. At no point should we draw a line and claim that we have certain knowledge in matters of fact. To repeat: if it isn't capable of being falsified, it isn't science.

SCIENCE AND PSEUDO-SCIENCE

If a person persists with the infuriating habit of claiming absolutely everything that he or she does as a great success, even if to the external observer it may appear a bit of a disaster, one might well ask 'What would have to happen for it to be described as a failure?' If absolutely nothing counts as a failure, then nothing merits the title 'success' either – both are rendered meaningless in terms of an objective or scientific approach, the claim of success simply reflecting the choice to see everything in that positive way. In other words, we need to recognize that for any claim to be accepted, it needs to be testable against experience.

The claim to be scientific rests on the methods used in setting up appropriate experiments or in gathering relevant evidence, and also on the willingness to submit the results to scrutiny, and to accept the possibility that they may be interpreted in more than one way.

The most obvious example of a pseudo-science is astrology. Astronomy is regarded as a science because it is based on observations, and any claims made today may need to be replaced owing to further observations in the future. Astrology, on the other hand, is not considered a science because it is based on a mythological scheme with an annual cycle of 'signs'. There is nothing that one might observe that could lead to the suggestion that Gemini should no longer rule those born in May; nor that the stars should be looked at in different ways, giving different star signs. Astrologers may be meticulous in their calculations and intuitively skilled in the application of their theories to individual situations. Astrology may even be shown to be of value to the people who practise it. But neither fact (if proved true) would even start to make astrology a science. For that to happen, it would be necessary to find evidence for an objective relationship between dates of birth and general behavioural tendencies; evidence which would then be open to scrutiny, and which could genuinely put the basis of astrology at risk.

Another example of pseudo-science is crystal therapy. The rationale given for having crystals about one's person, or under the pillow at night, is that they somehow have a 'vibration' that can influence moods. When challenged, the person might argue that having a vibration is not limited to crystals, but is a universal phenomenon. That may be fine as a scientific theory – it sounds scientific in that it uses language associated with science – yet this does not make any serious connection between general theories about atoms and their behaviour, and how you might feel calm and get a good night's sleep. Unless a theory for such a connection could be put forward in a way that was open to examination and testing, with the possibility that it might be proved wrong, then you do not have genuine science.

Insight

It's not that it's impossible for you to be affected by the date and time of your birth, nor is it absolutely impossible for the crystal beneath your pillow to induce calmness – it's just that those claims are made without allowing them to be scrutinized.

The issue about what constitutes science or pseudo-science is not always straightforward. Take the example of Marxism. Clearly, Marx's work on dialectical materialism was based on logic and the observation of the way in which societies in the past were organized and the factors that brought about change. In this sense, following the inductive method, Marx's theory might indeed be regarded as genuinely scientific. But the problem is that a Marxist may try to use that theory to interpret every event in such a way that, whatever happens, Marxism appears able to predict it. This creates a problem, which was at the heart of Karl Popper's criticism of both Marxist and Freudian thinking – a theory that is not open to be falsified cannot be scientific.

Insight
The distinction between science and pseudoscience is therefore essentially one of method and openness to criticism, rather than content.

Paradigms and 'normal science'

Thomas Kuhn (1922–96) struggled to understand how progress in science could be reconciled with either the idea of straightforward induction, or with the implications of Popper's falsification approach, where a single piece of contrary evidence was deemed sufficient to require the rejection of a theory.

He therefore developed an alternative view, based on an overview of how, from an historical perspective, science had gone about its business. He saw that there had been moments of insight when everything seemed to change as a result of a new theory, but that such moments stood out in contrast to a background of routine scientific research that built on and confirmed a set of already established ideas. Science did not simply get rid of theories and replace them every time a conflicting piece of evidence was found; rather – at least for most of the time – its work was gradual and cumulative.

Kuhn recognized that the basic set of assumptions that work for science over a particular period of time remain normative; that is, most scientists just get on with the job of carrying out experiments within a set of scientific assumptions that they have inherited. As laws and theories become established within the scientific community, they are used as a basis for further research; these he termed 'paradigms'.

He therefore distinguished between 'normal science' and those moments of crisis in which the whole approach is changed in what amounts to a scientific revolution. The periods of stability are dominated by a single paradigm, but over a period of time problems inevitably develop with the use of that paradigm and increase until they provoke a crisis, at which time an alternative paradigm may emerge, one that is able to deal with the problems that have caused the crisis. Once that new paradigm is accepted, science settles down once again.

This process by which paradigms are accepted during periods of 'normal science', and then set aside in crises, was set out in Kuhn's book *The Structure of Scientific Revolutions*, published in 1962.

Example

In cosmology, the most obvious examples of paradigms being replaced were the revolutions that allowed the world of Newtonian physics to replace the older Earth-centred world of Aristotle and Ptolemy, and that which came as a result of Einstein's theories of relativity, displacing Newtonian physics.

With hindsight we see the way in which philosophers and scientists sometimes affirm their particular vision of the world just as the scientific community is about to go through a 'paradigm shift' in which everything is reassessed. At the end of the nineteenth century nobody could have dreamed of the drastic changes that would happen in science during the first half of the twentieth, but at the time, given all they knew about the world, their confidence in the old paradigm made perfect sense.

A particularly controversial aspect of Kuhn's theory is that he claims that there is no independent evidence by which to decide between two different paradigms, since all evidence is either interpreted in the light of one or in the light of the other. We cannot make observations that are genuinely independent of the paradigm within which we operate, simply because those observations are shaped by the paradigm. Nor is it possible to have observations that are free from any paradigm. It is therefore possible to interpret Kuhn in such a way that he is seen as a relativist, since a paradigm can be evaluated against the questions asked by a particular society at a particular time, but cannot be compared with another paradigm from another period. In other words, each paradigm has its own language, with which the terms used by other paradigms may be incompatible.

Popper's theory of falsification might suggest that, if we find conflicting evidence, we should scrap our existing theory. In practice, however, that is seldom the case. Scientists become committed to the theories with which they work, and generally give them up only once evidence against them become significant or overwhelming. They work within a particular paradigm, and that influences their way of examining evidence and rationalizing from it. If experimental facts do not match the accepted theory, then the first reaction – before that theory is totally dismissed – is to see if there is some other factor influencing the anomalous results, or whether the existing paradigmatic theory needs some minor adjustment to accommodate the new evidence. An unexpected finding sends some scientists back to check their results and others to discuss whether adjustments to existing theories need to be made.

Insight

Paradigms are a common feature of human thinking. There are social, religious and cultural paradigms as well as scientific ones. Within any art, it is difficult to step outside one's tradition and produce something completely different. The baroque or classical style, for example, provided a paradigm within which composers or artists were content to work for most of their time. The triumph of imagination and creativity is to get beyond the paradigm that you have inherited.

There is some ambiguity about the use of the term 'paradigm', however. Sometimes Kuhn uses it to denote a single theory that is so successful that it becomes the exemplar for others; on other occasions 'paradigm' is used for the whole set of theories that together form a way of looking at reality. The term 'paradigm shift' for a major turning point in science clearly refers to the second of these meanings.

Progress

The implication of Kuhn's argument that there is no external, objective point of view from which to judge between paradigms is that we cannot show that science progresses. If things are lost as well as gained with a shift in paradigm, we may find, from some future perspective, that the shift was a retrograde step.

Kuhn's own answer to this is rather curious. From within a paradigm, because it took over from a previous one by appearing to address at least some of its shortcomings, its adoption is going to look like progress. But that view of progress is limited to the perspective given by the paradigm, and cannot claim to be objective. However, he also argues – for exactly the same reason – that, over time, there will be an increase in problem-solving, which would seem to be a valid and objective way of defining progress.

In *Against Method: Outline of an Anarchistic Theory of Knowledge* (1975), **Paul Feyerabend** (1924–94) argued that progress is misguided and impossible; we cannot achieve 'true knowledge', only various ways of seeing. Hence, for Feyerabend, the choice of one theory over another may be made for all sorts of personal, cultural, aesthetic and subjective reasons. Each person is free to choose his or her own view, and science cannot impose an absolute or fixed criteria for what is true and what is not – in other words, when it comes to the theory of knowledge, anything goes.

If that is the case, what is the motivation to do science? In the seventeenth and eighteenth centuries scientists thought that they were gradually dispelling ignorance and establishing the rule of reason. Can the same impetus be found if science is merely offering a succession of optional viewpoints? It may actually be the case, but can you seriously engage in science if you believe it to be so?

Insight

Feyerabend's position could not be more different from that of Popper. Popper has an ideal of truth or reality which, although never attainable, acts as a guide in the critical evaluation of theories. For Feyerabend, truth is related to one's personal commitment to a theory or paradigm. Kuhn hovers between these two positions, without objective criteria for assessing the relevant value of paradigms, but with an overall increase in problem-solving.

Notice also that the older inductivist approach to the use of evidence and experiment tended to suggest that progress was a very slow and cumulative business. With Kuhn, we see it quite otherwise: as an erratic progress, with moments of sudden advance separated by long periods of consolidation. Looking at the history of science, it is clear that Kuhn's is the more realistic view. An existing paradigm is going to be changed only if the evidence against it becomes overwhelming and an alternative is ready to take over.

But, clearly, there is some sort of progress, even if the paradigm is not changed in one of Kuhn's moments of revolution. One way of dealing with this is put forward by **Imre Lakatos** (1922–74) in his *Falsification and the Methodology of Scientific Research Programmes* (1970). Lakatos recognized that, in practice, science made progress by way of research programmes, which were essentially problem-solving activities. Within such a programme, one might distinguish between a 'hard core' of theories without which the programme would not be viable, and which scientists would not discard without overwhelmingly good reasons, and a 'protective belt' of supplementary theories which could be examined and changed without totally abandoning the overall programme. Hence progress is possible by constantly adjusting these less crucial theories.

In practice, there is likely to be more than one research programme on the go in a field of study at any one time. Progress can therefore be made when one of these is shown to be more fruitful than the others. Competition is not simply between theories, but between research programmes.

Lakatos thus criticizes Popper for not appreciating the historical continuity of theories within research programmes, making them vulnerable to falsification in a way that does not correspond to the actual way in which scientists evaluate their work. However, he also criticizes Kuhn for making changes in paradigm largely irrational affairs, produced by the choices of groups of scientists, without being able to specify why a new paradigm is superior to the one it replaces.

The status of scientific theories

If a theory is to gain acceptance, it is important that it should be compatible with other well-established theories. If predictions made by two theories are mutually exclusive, then one of them will eventually be shown to be unsuccessful.

Example

In the nineteenth century it was believed that the Sun generated its heat from the effect of gravity, crushing its mass together. In other words, the Sun was gradually shrinking, giving off heat and light as it did so. Various calculations were made about how long the Sun could go on shining, and thus about its age. Towards the end of the century Lord Kelvin (based on work done earlier by Helmholtz) came to the conclusion that the Sun and Earth must be about 24 million years old. This is sometimes referred to as the Kelvin–Helmholtz timescale.

But the problem with this was that, if Darwin's theory of natural selection was correct, the Earth needed to be far older. Both theories were carefully calculated: yet one had to be wrong. If Darwin was right about the time taken for species to evolve, then there had to be an alternative way in which the Sun could produce huge amounts of energy.

With Einstein's theory of relativity a few years later, that dilemma was resolved because it gave an alternative explanation for the long-term release of energy from the Sun. Of course, the Sun's fuel will not last forever but at least the theory of relativity gave a plausible explanation for it being old enough to have allowed time for evolution on Earth.

The key thing to remember is that the pictures or models by which we attempt to understand natural phenomena are not true or false, but adequate or inadequate. You can't make a direct comparison between the image and reality. Models operate as ways of conceptualizing those things that cannot be known by direct perception.

Hence the models by which we envisage those things that we cannot see directly need to be open to revision in just the same way as the theories within which they appear. In practice, one theory (or paradigm, even) seldom gives way immediately and obviously to another. There is frequently a period of overlap during which rival theories are compared. It is also common for a new theory to be dependent initially upon an older theory or paradigm, even if it subsequently becomes independent of it.

Copernicus' use of epicycles to explain planetary motion is an example of this reluctance or inability to break from an earlier model, even while proposing something radically new.

Thus, at any one time, scientists may be working with a number of different theories concerning any one particular area of research, one of which may come to be seen as more adequate or comprehensive than the others. There are times when a theory does not make great progress because other work alongside it needs to be done before its significance can be appreciated.

Example

Quantum theory was put forward by Max Planck in 1900, but its significance was not appreciated fully until after publication of the work of Einstein (from 1905) and Bohr (1913), since it seemed too much at odds with the physics of its day. But, of course, the difficulty of reconciling quantum mechanics and relativity is also an object lesson in the difficulty of comparing alternative approaches.

Naturally, the acceptance of a theory by the scientific community does not thereby guarantee its permanent status, but there are criteria by which one theory may be preferred to another. Acceptance may often depend on a theory's ability to predict successfully; the more its predictions are confirmed, the greater is its degree of acceptance.

At the same time, where there are equally successful theories, a choice between them may be made on the ground of simplicity or elegance. This follows from 'Ockham's Razor', a principle named after William of Ockham who argued that one should not multiply causes beyond necessity. In other words, if there are two competing theories, all other things being equal, the best policy is to accept the simpler.

In his book *The Essential Tension* (1977), Kuhn sets out what he saw as the five characteristics of a good scientific theory. They are:

1 accuracy
2 consistency
3 scope
4 simplicity
5 fruitfulness

He points out that these may well conflict with one another; there may be a conflict between (for example) a more accurate theory and one which, in practice, is more fruitful in enabling scientists to make more predictions.

Kuhn's main point (and a way of defending him against the criticism that his view of change is based on irrational factors) is that one scientist may prefer one theory because of certain qualities while another may, by placing emphasis on other qualities, favour another. He judges that, collectively, the scientific community comes to a common mind about which theories are better than others. This is not a simple matter of weighing evidence, but of taking all five factors into account.

THE DUHEM–QUINE APPROACH

So far we have looked at grounds for accepting or rejecting theories individually, or grouped within a paradigm, or as part of a research programme. There is, however, a line of argument that questions any attempt to divide our knowledge up in this way. It is generally known as a Duhem–Quine approach, after the physicist Pierre Duhem (1861–1916) and the philosopher W. V. Quine (1908–2000).

Duhem, writing in the 1890s, argued that you could disprove a theory only on the basis of other theories that you held to be true. If your existing and accepted theories were at fault, then your disproof of the new theory would be invalid. Thus, he argued that it was a mistake to try to separate theories off from one another at all; they should be taken all together as parts of a whole.

Quine took a similar line of approach in an important article, 'The Two Dogmas of Empiricism', written in 1951. He argued that our ideas fit together, so we should not assume that each and every statement can be reduced to sets of facts that can empirically confirmed in isolation:

> *Taken collectively, science has its double dependence upon language and experience; but this duality is not significantly traceable into the statements of science taken one by one… The unit of empirical significance is the whole of science.*

> From the end of section V of 'The Two Dogmas of Empiricism'

Quine recognized that science is a human construct, and not something that can simply be reduced to a catalogue of bits of empirical evidence:

> **The totality of our so-called knowledge or beliefs, from the most casual matters of geography and history to the profoundest laws of atomic physics or even of pure mathematics and logic, is a man-made fabric which impinges on experience only along the edges.**
>
> Opening of section VI, 'The Two Dogmas of Empiricism'

In other words, Quine makes the point that you can't simply pick off individual bits of knowledge and expect to justify them from experience. Science holds together as a 'fabric' and the status of individual theories should be seen in the context of the whole.

....

Insight

In his article, the two dogmas attacked by Quine were:

1 the traditional distinction made between **'analytic' statements**, which were deemed to be true or false depending on their internal logic independent of facts, and **'synthetic' statements**, which referred to matters of fact and could be judged on the basis of evidence. (Quine held that the terms used in analytic statements had a meaning that depended upon their use, and that depended upon facts.)

2 the idea that every meaningful statement is related to matters of immediate experience (which is what the Logical Positivists had argued).

It is this second dogma that led Quine to the idea that meaning should be related to the whole of science, rather than to each individual statement.
....

Working to understand some unusual evidence, or the unexpected results of an experiment, one may suddenly come up with a new theory to explain them. A key question to ask at that point is: 'How does this new theory relate to everything else I believe to be true? What other theories have to be changed to accommodate this new one?' In a sense, the Duhem–Quine approach is the recognition that you cannot make a move on the chessboard without influencing the direction of the whole game. Theories have to hang together, or they become meaningless.

KEEP IN MIND

1 Logical Positivists and others sought empirical confirmation and reduced everything to sense experience.

2 Popper argued that, to be scientific, a theory must be capable of falsification.

3 For Popper, truth is an ideal towards which science strives through critical engagement with existing theories.

4 Kuhn contrasts moments of crisis with 'normal science'.

5 Kuhn also argues that everything is to be understood from within a paradigm.

6 Feyerabend advocates anarchy in the theory of knowledge, with no objective criteria for deciding between theories.

7 Lakatos sees progress through competing research programmes.

8 A model should be judged by whether it is 'adequate', rather than 'true'.

9 Kuhn's five criteria for a good theory: accuracy, consistency, scope, simplicity and fruitfulness.

10 Quine sees our ideas as fitting together like a fabric.

Realism, language and reductionism

In this chapter you will:
- *consider whether science can describe the 'real' world*
- *explore how language is used in science*
- *examine the implications of a 'reductionist' viewpoint.*

'Scientific realism' is the term used for the view that the objects with which science deals are separate from, and independent of, our own minds, and that scientific theories are therefore literal descriptions (whether true or false) of the external, objective world. The implication is that we can assume that there is a truth out there to be had, even if we have not yet found the perfect theory by which to describe it.

Clearly, most people assume that this is the case. The whole point of the development of scientific method – setting up experiments and gathering impartial evidence – was aimed at achieving knowledge that was free from the influence of personal interests or received tradition. However, in looking at ways in which theories and paradigms are developed and replaced, it is clear that things are not always straightforward. We may interpret evidence in the light of existing theories and there are areas of science (e.g. particle physics) where it is acknowledged that the act of investigation itself influences what is investigated. How can that yield knowledge that is truly independent of our own minds?

If you look back on the history of science, it is clear that most theories are eventually discarded and replaced. Hence it is wise to assume that even the best of today's theories will eventually need to be discarded, and therefore that we cannot assume that they give a literal or true picture of reality. That would suggest that we should take an anti-realist position, considering theories to be ways of summarizing our own experience, rather than giving a literal picture of reality.

Add to that the fact that competing theories are often underdetermined (in other words, there is insufficient evidence to decide between them) and – certainly if Feyerabend is right – we may choose one theory rather than another for all kinds of reasons unrelated to their factual accuracy. We may try to find some other way to decide between theories – for example that one is simpler, or more useful for prediction. After all, Newtonian physics was regarded as the final word because it was so good at predicting. Equally, we might follow Quine and explore how our theory fits the whole fabric of science, testing it out against other things that we hold to be true. All of these options may be useful for the practice of science, but they do not suggest that one theory represents a literal, true picture of reality and that the others must therefore be wrong. Of course, following Popper, we may hope that we are gradually getting closer to the truth, but that is an assumption or a hope; it is not something that can be proved.

Insight

Do you believe that it is possible to have a perfect theory? One that literally describes reality at a level beyond that which you can observe? If you do, you are taking a realist view. If, on the other hand, you think that all theories are ways of making sense of our experience, but cannot describe reality as it is in itself, you are taking a non-realist position. Thus, for example, in the debate between Einstein and Bohr, Einstein held a realist position, while Bohr held to the view that, at the quantum level, theories were non-realist and assessed on a pragmatic basis.

Reality and observation

Many apparently 'modern' issues can be traced back to the philosophy of ancient Greece, and the issue of scientific realism is no exception. In the Pre-Socratic times there was a fundamental disagreement between Protagoras and Democritus. Protagoras argued that all we could know were the sensations we received. We could know nothing of what was out there causing those sensations.

Consider the statement 'I see a red ball'. That is true or false depending on whether my eyes have indeed recorded light of that wavelength forming a circular pattern on my retina. I have to infer that something outside my eye has caused that to happen, but all I actually know is that optical phenomenon. (There is an additional problem concerning interpretations. I may think that I am looking at a red ball, but on approaching it I may see that it is, in fact, a red apple. The sensations remain the same initially, but my mind interprets them as one thing or the other.)

By contrast, Democritus insisted that 'things' existed independently of our perception of them. That is an equally logical view, since the red ball surely continues to exist if I shut my eyes. Confronted with the same question, the eighteenth-century philosopher George Berkeley (1685–1753) argued that, since we only know things exist when we perceive them, they can only continue to exist if perceived by God.

Example

If simple observation were the only factor in determining our knowledge of reality, then nothing at all can be more certain than the fact that the Earth is stationary. For thousands of years humankind has observed the turning of the stars and (apart from the odd earthquake) has experience the ground beneath its feet as a fixed point from which to observe all other movement. To accept that the Earth moves around the Sun and turns on its axis on a daily basis is to move away from simple experience, to start to interpret what is seen in the light of a theory. Against all the evidence of our senses, we 'know' that we are hurtling through space. Uninterpreted evidence is therefore an inadequate basis for any scientific theory.

The key point here is the distinction between what actually exists (**ontology** – the theory of what exists) and what we can know about what exists (**epistemology** – the theory of how we know things). Ontologically, it makes sense to say that 'things' have an existence independent of our perception of them. Epistemologically, it makes sense to say that we cannot know things except by our perception of them.

The actual process of observation is complex. The idea of space and the distance between objects relies on the brain linking one thing to another; the conventional idea of time appears as we remember that some experiences have already taken place. If science depends on experience, then it is dependent upon our ways of looking, thinking, recognizing and remembering. The matter becomes more complex when we consider how modern science makes connections between what we actually see or produce in the course of an experiment and what we infer to be objectively the case.

Example

When a metal is heated to very high temperatures in an electric arc, it emits light which can be represented as a spectrum of lines. The pattern and sequence of those lines is always the same, wherever, for example, iron is present. Of course, this applies only to iron that is hot enough to vaporize; at lower temperatures it glows and emits light across a continuous spectrum, and the lines disappear.

This means that we are able to detect metals across huge distances. We can know that a particular metal is present in a distant star, without being there to analyse it. All we need to do

> is look at the spectral lines produced by the light from
> that star. If the lines match those of light given off by iron
> vaporized here on Earth, the conclusion is that iron is also
> present in that star.

In other words, what we perceive tells us about what is 'out there', but it not identical to what is 'out there'. Reality is inferred from observation, not identical with it. In this case, the presence of a vaporized metal is the best explanation for the sequence of lines in the light from the star.

Clearly, our sensations are limited to objects within a narrow size and distance range. Most of what we know in science, therefore, we have to infer from the best explanation that science offers us for those perceptions. This process, which is central to the way in which science works, is termed 'inference to the best explanation', a form of abductive reasoning (see Chapter 3).

Inference of this sort is neither new nor unreasonable. From early times, humans saw a movement of leaves and inferred the presence of an animal in the jungle. One key difference was that primitive humans did not stop to contemplate whether the inferred animal was real or not. Depending on its size, they would either have needed to kill and eat it, or retreat quickly before it killed and ate them! Their skills at observation and inference were judged on a pragmatic basis. Science can take a similar approach: if an inferred theory works and is useful, it is provisionally accepted.

Insight

The big issue here is whether you accept scientific realism (that our theories can describe real unobservables) or accept some form of empiricism (that we can do no more than account for the phenomena we observe). Some empiricists argue that the test of a theory should be that it is empirically adequate – in other words, that it is correct in what is says about those things that we can observe, even if it cannot tell us about those things that are unobservable.

OBSERVATION IN QUANTUM THEORY

According to the 'Copenhagen interpretation' of quantum theory (so called because it was developed at the Copenhagen Institute of

Theoretical Physics), particular states only become determinate when we observe them. In other words, our act of observing brings reality into being. Thus, according to quantum theory, everything is actually interrelated and nothing is determined. But in an observed universe, by the act of observation, everything is determined.

In terms of scientific realism, there was a fundamental disagreement about quantum theory between Bohr and Einstein.

▶ For Bohr (and Heisenberg, who worked with him), the uncertainty that applies to subatomic particles is not just a feature of our observation, but is a fundamental feature of reality itself. It is not just that we cannot simultaneously know the position and momentum of a particle, but that the particle does not have these two qualities simultaneously. What is there is what we perceive; our observation creates the reality of what we are observing.

▶ Einstein, on the other hand, took the view that there was indeed a reality that existed prior to our observation of it. Thus, a particle would have a position and a momentum at any instant; the only thing was that it was impossible for us to know both at the same time. Reality is thus prior to observation. But, of course, it remains essentially an unknown reality, since, as soon as we try to observe it, we are back in Bohr's world where what is there is determined by our observation.

SCHRÖDINGER'S CAT

A well-known but often misinterpreted example of the issue of whether uncertainty is a feature of reality itself or merely of our observation is known as 'Schrödinger's Cat'. Schrödinger was opposed to Bohr's Copenhagen interpretation of quantum theory. He illustrated his point by suggesting this thought experiment:

Suppose one were to put a cat into a sealed box, along with a bottle of cyanide, which will be smashed by a hammer if there is any decay of a radioactive substance within the box. Is the cat alive or dead?

According to Schrödinger (and common sense), the cat is either alive or dead in the box. We cannot know which is the case without opening it and taking a look – nevertheless, the reality is that the cat is either alive or dead prior to our opening the box. According to the Copenhagen interpretation, however, the cat is neither alive nor dead until we open the box!

Thus, Schrödinger takes a view that follows Einstein's criticism of Bohr – that uncertainly is a feature of our observation, not of reality. One way of describing the situation is to say that Bohr and Heisenberg argue for essential indeterminism, whereas Schrödinger and Einstein accept only an indeterminism of observation.

Insight

These arguments should not be applied at an inappropriate level. According to Schrödinger, we cannot apply the principles of quantum physics to human beings; they operate at very different levels of reality. We do not experience ourselves as being determined in the process of being observed – unless, perhaps, in terms of being obliged to adopt particular social roles and understood in terms of them!

GENERALIZATIONS

One argument in favour of the anti-realist view of scientific theories concerns the very nature of what a theory is. Theories are generalizations; they attempt to show and to predict across a wide range of actual situations. Indeed, the experimental nature of most scientific research aims at eliminating irrelevant or particular factors in order to be able to develop the most general theory possible.

Now in the real world (as was pointed out by Duhem and others) there are no generalities. You cannot isolate an atom from its surroundings and form a theory about it. Everything interconnects with everything else – all we have are a very large number of actual, unique situations. Our theories can never fully represent any one of these because they try to extract only generalised features.

THE 'MIRACLE' ARGUMENTS

Miracles don't feature much in the philosophy of science, but it is worth considering two arguments connected with them, one from the eighteenth century and the other an argument in favour of scientific realism.

The first comes from David Hume. It starts by making three assumptions:

1 A 'miracle' is an event that violates a law of nature. (That is not necessarily the view of miracles that some religious people today would take, but it is sufficient for the purposes of this argument.)
2 Laws of nature are established on the basis of the best and most extensive evidence available.
3 A wise person proportions belief to evidence.

In which case, Hume argues that there can never be sufficient evidence to prove a miracle, since the evidence against a miracle (i.e. for the law of nature) is always going to be greater than the evidence that a miracle has actually happened. This argument does not claim that a miracle (on his terms) cannot happen, simply that we can never prove that it has happened on the basis of evidence.

With that argument in mind, we can turn to the 'miracle argument' in favour of scientific realism:

> If a theory is successful (i.e. that we can use it to make accurate predictions), which of the following conclusions is more likely?
>
> ▶ that the theory is successful because it is 'true' in the sense that it reflects reality
>
> ▶ that the theory is false; that it does not, or cannot, reflect reality but just happens to work successfully.

The 'miracle argument' claims that it would be a miracle if a theory that did not reflect reality just happened to be successful. Like David Hume's argument, this cannot prove that a theory reflects reality, simply that it is the best explanation for its success.

And, of course, the anti-realism position does not try to claim that a theory is wrong, simply that its value is found in the predictions it is able to make; not so much a matter of right or wrong, but of being successful or unsuccessful.

Language

Ludwig Wittgenstein starts his book *Tractatus Logico-Philosophicus* with a remarkable statement:

The world is everything that is the case.

Tractatus 1

He took the view that the function of language was to picture the world:

The totality of true propositions is the whole of natural science

Tractatus 4.11

He ends the work with the equally famous phrase:

Whereof we cannot speak, thereof we must remain silent.

Tractatus 7

Now, Tractatus has been a hugely influential book, and – as is well known – later in his life Wittgenstein was to develop a very different approach to language. But for our purposes here, apart from the fact that he saw the function of language as pointing to external facts, we need to note one absolutely crucial thing: that science is to do with propositions, not with external 'things'.

Science does not make the atom what it is, nor create DNA, nor shape the universe from the 'Big Bang' onwards. Science is not the same thing as the world it investigates. It is – exactly as Wittgenstein said – the totality of propositions. (I omit his reference to 'true' because many scientific propositions are known to be false, or may one day be shown to be false.) But the fact remains that 'science' is a network of words, ideas, mathematical calculations, formulae and theories. It is a form of language. It is a human construct. That is why it is both possible and important to have a philosophy of science, for once investigations and experiments are carried out, their results are evaluated and find their place within this ever-changing network of propositions. Without thought and language, science does not exist. Facts are the product of a thinking mind encountering external evidence, and they therefore contain both that evidence and the mental framework by means of which it has been apprehended, and through which it is articulated.

We all have conceptual frameworks which suggest to us how we should interpret our experience. We never come to a new experience without some sort of anticipation about what it will be like. The words we use to describe it are part of that framework. They relate this new experience to what we and other people have known in the past. They are a kind of shorthand that saves us from having to begin from scratch to describe the elements of everything we see.

So the words we use, and the facts that they express, have meanings that are already given to us by the society that shares our language. You cannot describe something unless you can find existing words that convey something similar and, whether those words are used literally or metaphorically, they colour and give meaning to what is described. Therefore, a proposition is never neutral in terms of language; it is never free from all that language has become as it has developed and grown its vocabulary. Everything we see, we see 'as' something; that is a general feature of experience, but it is also the result of using language.

Language is active in shaping our experience; it is not simply a transparent medium through which experience may be communicated. The propositions of science are not derived solely from facts, although they often appear to be so. Propositions always depend on other propositions and are therefore always open to question, always fallible. We can understand a scientific theory only because the words and ideas in which it is expressed are already familiar to us; if we misunderstand the words, we fail to understand the theory.

CLARITY AND CORRESPONDENCE

Clarity is a key feature in any language that is to convey scientific theories accurately. A good example of the scientific insistence on correct language is in the complaints made by Galileo about those who tried to cover over their ignorance about causation by saying that something 'influenced' or 'had an affinity with' something else. These, he believed, had no factual meaning, and it would have been far more honest to say that one did not know any way in which the one could cause the other. He considered that all such vague use of language gave the impression that the speaker was pretending to offer a reason, but without any concrete evidence, and therefore had a tendency to mislead.

What then can be said about the world, and what cannot? We have already encountered the Logical Positivists. Impressed with the obvious success of the scientific method, they sought to accept as factually correct only those statements whose meaning could be verified with reference to the sort of evidence that would be appropriate in a scientific experiment. Following the work of Mach and Russell, they divided all statements up into logical and mathematical terms on one side and claims about empirical facts on the other. The former were known prior to experience (a priori) but the latter needed to picture the objects of sense experience.

The clear implication of the whole approach of the Logical Positivists was that the language of science should simply offer a convenient summary of what could be investigated directly. In other words, a scientific theory is simply a convenient way of saying that, if you observe a 'particular' thing on every occasion that these 'particular' conditions occur, then you will observe this 'particular' phenomenon. What scientific statements do is to replace the 'particulars' of experience with a general summary.

Clearly, the ultimate test of a statement is therefore the experimental evidence upon which it is based. Words have to correspond to external experienced facts.

The problem is that you can say, for example, that a mass or a force is measured in a particular way. That measurement gives justification for saying that the terms 'mass' or 'force' have meaning. But clearly you cannot go out and measure the mass of everything, or the totality of forces that are operating in the universe. Hence such general terms always go beyond the totality of actual observations on which they are based. They sum up and predict observations.

For the Logical Positivists, the truth of a statement depended in being able (at least in theory) to check it against evidence for the physical reality to which it corresponds. Scientific theories can never be fully checked out in this way, since they always go beyond the evidence; that is their purpose, to give general statements that go beyond what can be said through description.

The problem was to find the correspondence rules by which one could relate a particular term to the observations upon which it was based. If you couldn't, in theory at least, specify the observations upon which something was based, then it was meaningless.

Now clearly, if you could observe something directly, there would be no problem. The task of scientific language is to describe what goes beyond observation. The question is how it can be shown to do so, or to what extent language is actually selected and given its validity by the subjective wishes of the person using it, or by the society that creates its meanings.

> **Insight**
> Where science gives a description of a phenomenon, the simplicity of pictorial language may be adequate. But what of its explanations, inferences and hypotheses? These are not statements that can easily be reduced to a sequence of words each of which corresponds to a piece of empirical evidence. Both science and language are far more flexible and interlocking than the Logical Positivists assumed.

However, the legacy of positivism is that scientists are rightly wary of statements that are not supported by empirical facts and of making unwarranted jumps from evidence to what 'really exists'. This is illustrated by the following quotation from Stephen Hawking in *The Universe in a Nutshell* (2001), which also endorses an instrumentalist view of theory, in his discussion of the idea that there may be ten or eleven dimensions of spacetime:

> *I must say that personally, I have been reluctant to believe in extra dimensions. But as I am a positivist, the question 'Do extra dimensions really exist?' has no meaning. All one can ask is whether mathematical models with extra dimensions provide a good description of the universe.*
>
> Stephen Hawking, *The Universe in a Nutshell*, page 54

DISPOSITIONAL PROPERTIES

When a material is described, it is necessary to say more than what it looks like; one needs to say (based on experiments) how it is likely to behave in particular circumstances. Thus, if I pick up a delicate piece of glassware, I know that it has the dispositional property to be fragile. The only meaning I can give to that property is that, were I to drop the glassware, it would break.

Now the term 'fragile' is not a physical thing; it is an adjective rather than a noun. I cannot point and say 'there is fragile'. I use the word 'fragile' as a convenient way of summarizing the experience of seeing a variety of things dropped or otherwise damaged. Those that are easily broken are termed 'fragile'.

The curious thing about the dispositional property 'fragile' is that it passes the test set by Logical Positivism (that the meaning of a statement is its method of verification), but only retrospectively. It would be easy (but expensive!) to verify that all glassware is fragile, and once verified, the thing verified would no longer apply, since we can say of a broken piece of glassware only that it 'was' fragile.

REDUCTIONIST LANGUAGE

We saw earlier that Wittgenstein (in the *Tractatus* – his views were later to change) and the Logical Positivists aimed to assess all language in terms of the external reality to which it pointed, and to judge it meaningful if, and only if, it could be verified by reference to some experience. If a statement could not be verified (at least in theory), then it was meaningless. The only exceptions to this were analytic statements, which were true by definition.

The Logical Positivists therefore believed that all 'synthetic' statements (i.e. those true with reference to matters of fact, rather than matters of definition) could be 'reduced' to basic statements about sense experience. However complex a statement may be, in the end it comes down to pieces of sense data, strung together with connectives. The ability to reduce language in this way was one of the 'dogmas of empiricism' attacked by Quine (see Chapter 4).

Such reductionism is primarily about language, but how we deal with language reflects our understanding of reality. So reductionism influences the approach that we take to complex entities and events, and it is to this broader sense of reductionism that we now turn.

Reductionism and its implications

When we examine something, there are two ways we can set about doing so:

1 A **reductionist** approach is to take it apart to see what it's made of. The 'reality' of a complex entity is found in the smallest component parts of which it is made.
2 An **holistic** view examines the reality of the complex entity itself, in the light of which its individual component parts are subsequently understood.

Science can operate in both ways. On the one hand, it can analyse complex entities into their constitutive parts, and on the other, it can explore ways in which individual things work together in ways that depend on their complex patterning.

The other side of the reductionist phenomenon is 'emergentism', in that, with every new level of complexity, features emerge that are simply not visible at the lower level. Hence the need (in addition to reductionist analysis – these approaches are not mutually exclusive) to consider complex entities from a holistic perspective.

Consider the human body. At one level, I am nothing more than the sum total of all the cells of which my body is made. But those living cells can be 'reduced' to the chemical compounds of which they are made, and then further 'reduced' to the constituent atoms, which follow the laws of physics. By taking this approach, although the human body is complex, its operations can ultimately be analysed in terms of the laws of physics. But does the operation of each atom, within each molecule, within each cell, within my body, explain who I am?

Nobody would deny that human beings are composed of atoms, what can be denied is that a human being can be adequately explained in terms of them. From the holistic point of view, a human being has a personal life that is quite different from those of the individual cells of which it is made up, and operates at a different level.

Example

No amount of analysis of the effects of solar radiation on individual tissues, or the protective qualities of melanin, is going to explain why people choose to take a holiday and lie on a beach!

Sunburn can ruin that holiday; but your social life and the wellbeing of individual cells on the surface of your skin – although they can sometimes impinge on one another most intimately – are operating at very different levels. However it may feel at the time, there is more to you than sunburn!

Clearly, therefore, the two ways of considering human beings are not mutually exclusive. Whether it is mild sunburn, or a serious cancer, the behaviour of individual cells is significant for the operation of the whole body. Equally, a holistic event (e.g. getting excited or worried about something) has an immediate effect on many of our bodily systems.

It does mean, however, that a reductive analysis is not the only scientific approach to understanding human beings, any more than a sophisticated computer program is completely explicable in terms of the binary code of which it is, in the final analysis, composed.

Within the overall scope of science, reductionism has been fundamentally important – we often understand things best once we have taken them apart – but it may not be the best way of appreciating the emergent properties of complex entities.

KEEP IN MIND...

1 Basic question: Do we know things, or only our experience of them?

2 Realism is the claim that unseen entities in science actually exist.

3 Reality may be inferred from observation, not identified with it.

4 In the Copenhagen approach to quantum theory, observation determines what is seen.

5 Einstein took the view that uncertainty is a feature of our observation, not of reality.

6 Language shapes experience.

7 Interpretation is part of the process of observation.

8 Reductionism in language sees meaning in the verification of individual statements by means of experience.

9 Reductionism tends to see reality in terms of the smallest level of its constituent parts.

10 An holistic view considers properties that emerge with higher levels of complexity.

6

....

Relativism, instrumentalism and relevance

In this chapter you will:
- *explore the idea that the theories we accept influence what we observe*
- *consider whether theories should be judged in terms of what they achieve*
- *examine issues about the relevance of particular theories within the scientific community.*

From time to time philosophers have sought a point of absolute certainty from which to build up the edifice of human knowledge. Probably the most famous example was Descartes, who was determined to doubt everything possible in order to find the one indubitable statement (which was, of course, his famous 'I think, therefore I am'). A craving for clear-cut, unambiguous language can be found in David Hume, for example, who wanted to cut away all metaphysical nonsense and base his understanding on sense experience. The Logical Positivists, too, wanted a form of language that would allow no scope for metaphysics, nor for anything that could not be evidenced through the senses.

Sadly perhaps, we have already seen that such a hope of certainty is illusory, for phenomena may generally be interpreted in a variety of ways. Kuhn argued that normal science works within a particular paradigm, which could then be replaced by a different one without there being any objective way to decide between them; Lakatos saw research programmes competing with one another, and Feyerabend argued that, effectively, anything goes when it comes to accepting theories. Such views come under the general heading of 'relativism', and it poses a major challenge for scientific realism

In this chapter, we shall explore these issues further by looking first at the theory-laden nature of all scientific observations, which strengthens the case for some measure of relativism. Then we move on to 'instrumentalism' and to 'relevance' as ways of addressing the key question: How can I decide between alternative theories?

Theory-laden observations

We saw in Chapter 3 that the process of induction is based on the idea that it is possible to get information about the world which is independent of the person who gathers it. Francis Bacon insisted that a scientist should set aside all personal preferences in assessing data, in the hope that any theory derived from it should equally apply to other observers and other times.

That was the ideal of the inductive method of reasoning. In practice, however, it is clear that when we make observations we already have in our minds various theories and ideas that may influence both what we see and how we select what is worth examining.

Sometimes, observations in one area of experience can lead to a general theory in another. In *On the Origin of Species*, for example, before putting forward his theory of natural selection, Darwin considered the way in which people breed animals and plants, interfering in the normal reproductive process in order to enhance or eliminate certain features. Through these observations, he was led to ask what factors in the wild might operate to bring about the same progressive changes that human breeders are able to generate artificially. He observed the competitive struggle for finite resources, where those best adapted to their environment were able to survive and breed. And from that observation he developed the theory of natural selection.

Did Darwin have the model of domestic selective breeding in mind while gathering the evidence that led him to formulate his theory of natural selection? If so, was it conscious? And, whether conscious or not, did it influence the way in which he observed, gathered and presented evidence for his theory? Looking at *On the Origin of Species*, we can see the way in which Darwin moves from observations, to the framing of models to account for them, to an overall theory. We can also trace ideas that influenced him (e.g. Malthus and Lamarck). The overall effect is to recognize that Darwin did not come to observe creatures on the Galapagos Islands with a completely blank mind, devoid of theories. As he observed them, he must have been constantly sifting through his fund of existing knowledge for some explanation for what he saw. The detailed records he made showed what he considered to be significant. The brilliance of Darwin's mind is seen in the way in which he pulls together existing ideas and welds them into a radical, new, inclusive theory.

Insight

The general point that this illustrates is of fundamental importance for science. Observations and evidence are not free from the influence of theories. We examine things with some purpose and therefore with some idea in mind.

Clearly, the scientist (or anyone else, for that matter) cannot be a sufficiently detached observer that his or her habitual way of looking at the world does not influence how things are seen – we have already seen this in Kuhn's account of 'normal science' where work is done within an accepted paradigm. Kuhn's relativism, at its simplest, amounts to the acknowledgement that we always have to opt for some frame of reference, and that, once we choose that frame of reference, it will influence how we interpret what we see. The important thing is to be aware of such influences and take them into account.

INTERPRETATION

Paul Feyerabend pointed out that interpretation is an integral part of the process of observation. Whatever I come across, I interpret, and all such interpretation is based on my previous experience.

Example

I look into the sky and see a small black dot. The fact that I see that dot is neither true nor false; it is simply a fact. If I go on to say 'There's a plane' that might be true; on the other hand, if the dot turns out to be a bird, then my statement is false. Observation is sense data plus interpretation.

The Logical Positivists had argued that, in order to show that something is correct, I must be able to show ways in which that statement corresponds to external reality. I need to specify my evidence. Feyerabend argued that this is, in principle, impossible, since the evidence I produce is again part of my interpretation of the world. I cannot get outside that interpretation.

The result of this is that one cannot 'fit' one's statements and interpretations to the world itself. One cannot say that they are either true or false in an objective way; so, for Feyerabend, there is no 'truth' in science.

Different people have different ways of interpreting experience. Each has an overall *Weltanschauung* (world-view). The problem is that there is no way of judging between different world-views, there is no way of getting beyond them and comparing them with some objective (uninterpreted) reality.

Insight

This dilemma of not being able to get outside one's own interpretation reflects the fundamental issue, presented by Kant in the eighteenth century, in distinguishing between things in themselves (**noumena**) and things as we perceive them to be (**phenomena**). The problem is that we can't get around phenomena or achieve some independent knowledge of reality. Things in themselves remain unknowable; all we have are our perceptions of them, shaped by our senses and the way our minds work. Nor can we get a view of the world that is from nowhere, since every view is from somewhere and that 'somewhere' determines how we see the world.

Of course, scientists do not work alone. Karl Popper argued that science was not subjective, in the sense of being the product of a single human mind, but neither could it be strictly speaking objective, since it is not made up of uninterpreted experience. Rather, he argued, science transcends the ideas of individuals and is to be understood mainly in terms of the whole scientific community.

A similar idea is found in the work of Lakatos, who speaks of theories being developed within 'research projects' rather than in isolation from one another, and Kuhn, for whom 'normal science' is carried on within an existing framework of ideas. Theories appear to have a measure of objectivity simply by the fact that they are shared within the scientific community, rather than being the product of a single individual. They can be judged by how well they fit with other established theories.

ALTERNATIVE MODELS

Earlier, when looking at falsification, we saw that more sophisticated approaches to falsification allowed that a theory would be discarded only when an alternative was found that could account for whatever evidence suggested that the first theory was inadequate. In other words, a process of scientific enquiry should be constantly looking to see if there are alternative theories to account for the evidence to hand – the existing theory is then rejected if another emerges that is more comprehensive.

> ### Example
>
> There is nothing wrong with Newtonian physics if your interests are limited to basic mechanical devices on the surface of this planet. On the other hand, in extreme situations, one needs to move from Newton to Einstein. Newtonian physics is therefore seen as limited rather than wrong. Einstein's theories of relativity are preferred simply because they can predict events that are beyond the scope of Newtonian physics.

Radically different theories may be equally compatible with the evidence. How then do you choose between them? One criterion might be the degree to which they can 'fit' existing evidence and theories. Thus, for Kuhn, during periods of 'normal science' new theories tend to 'fit' the existing paradigm.

Alongside conformity to existing theories, a new theory may be examined in terms of its inherent simplicity and elegance. As we saw above (Chapter 4), Kuhn thinks that there are five different qualities to be taken into consideration.

Yet this should not deflect us from recognizing the serious problems caused by the variety of possible interpretations. This is the issue of **underdetermination**. Quine argued that theories are underdetermined when there is insufficient data to be able to decide which of the available theories is best able to account for it, and therefore that is it possible to hold whichever of them you choose. The logic here leads in the direct of Feyerabend's 'anything goes' view, and is a major factor in supporting a relativist approach to scientific theories.

Insight

There is an interesting parallel here between the work of philosophers of science in looking at the variety of possible theories (e.g. Quine, Feyerabend) and that of philosophers of language who undertake deconstruction (e.g. Derrida). In literature, this led to the view that a text could have any number of different interpretations, and that there was no definite 'meaning'. A text could mean whatever one chose it to mean, with no ultimate criterion of interpretation.

Some authors and literary theorists argue that this produces anarchy and chaos, and is not in line with the intention of the author who produces the text in the first place. In the world of science, a similar complaint can be made by those who favour a 'realist' view of science – namely, that there needs to be some way of evaluating competing theories if there is to be progress towards understanding the nature of reality itself.

Lakatos argued that the history of science shows that different 'research programmes' work alongside one another in a competitive way, and it may therefore be far from clear which of them will come to dominate. He thought that it is right to work with a new theory, giving it the benefit of the doubt and shielding it from attack from more established theories, even if it is yet to yield any positive results. He took this position because otherwise he saw little practical opportunity for science to make progress, since a majority of new possibilities would be snuffed out before they ever became sufficiently established to mount a serious challenge to an existing theory or paradigm. In practice, it takes a really crucial experiment to provide sufficient evidence to establish or refute a theory, and constructing such an experiment is a far from obvious task.

In practice, a research programme should yield both new facts, but also new theories; in other words, it should produce what amounts to a programme of continuous growth. It is not enough for a theory to have a fundamental unity and consistency. Thus, for example, Lakatos is critical of Marxism as a theory, arguing that, since 1917, Marxism had not produced any new 'facts'. In practice, he was therefore critical of any theory that had become monolithic and static. He pointed out that:

The direction of science is determined primarily by human creative imagination and not by the universe of facts which surrounds us.

From his collected papers, published posthumously in 1978

Instrumentalism

Another of the qualities that Kuhn suggests we look for in a good scientific theory is **fruitfulness** – the degree to which each theory is able to make accurate predictions, in other words, its usefulness as an instrument for the application of science. He argued that the main test of science is its ability to solve puzzles: if a theory does this, it is useful and is developed; if it fails, then scientists look elsewhere.

Therefore, even if we cannot decide if a theory is right in any absolute sense, we can at least check to see what it can do for us. Indeed, we saw in the section on the inductive method (see Chapter 3) that making and checking out predictions based on a theory is a standard way by which that theory is assessed scientifically. Theories are therefore to be evaluated as tools for predicting; a view generally known as **instrumentalism**.

Science generally tends to take a pragmatic rather than an absolute approach to truth. We cannot be certain that any of our theories are absolutely correct, and we know that they will in all probability be replaced eventually by better ones (which in itself is an argument in favour of relativity). Therefore, in practice, we have to go for the most useful way of understanding the world that we have to hand, even if the theories we use already have recognizable limitations. We cannot wait for perfection.

A safer, instrumentalist bet

Copernicus' Sun-centred cosmology was presented, in the preface to his work by the theologian Osiander, as a useful tool for calculating the motion of the planets, rather than a way of describing what was actually the case. That was exactly what we would now call an 'instrumentalist' approach.

Notice that this approach gives scientists a very active role in seeking out theories and evaluating them. It is not a matter of waiting for evidence for a theory to be overwhelming before it is accepted, but of seeking out the theory that is most useful. The nineteenth-century French mathematician, physicist and philosopher Henri Poincaré (1854–1912) saw the mind as playing an active role, building up schemes by which we can understand the evidence we receive, and Duhem, at the beginning of the twentieth century, assessed theories in instrumental terms, looking at their explanatory power. Yet he believed that, as we construct concepts by which to grasp reality, we still hope that, in the end, we will find a theory that is neutral and will reflect reality as it is in itself.

Thus, even if our present assessment of theories is instrumental, we can retain a goal of some perfect theory that will eventually explain reality as it is.

Karl Popper argued that, when you have a choice of theories to explain some phenomenon, you should opt for the one that is better corroborated than the others, is more testable, and entails more true statements. In other words, you should go for the one that is most useful to your task in hand. And he says that you should do this, even if you know that the theory is actually false!

But we now need to step back from that pragmatic use of theories and ask about the implications it has for their status and function. What is clear is that theories are not 'out there' waiting to be discovered; they are instruments we devise in order to understand the world and are judged according to how effective they are in helping us to do so. Popper called theories 'instruments of thought'. They are our own inventions; highly informed guesses at explaining phenomena.

Insight

No theory is perfect, and in a world of competing research programmes, theories too will compete with one another. Truth may remain the absolute goal, but in practice it is the instrumental value of theories that determines which are accepted and which are rejected.

Theories may be accepted on the basis of their success in making predictions, but we can then ask: Why should a theory be successful in this way? One possible answer is that it succeeds in its predictions simply because it gives a true description of reality, which is the conclusion from the 'miracle' argument for scientific realism. This would suggest that instrumentalism, far from tending to promote relativity between theories, actually supports the quest for realism, by favouring those theories that, by their success, suggest that they reflect more accurately what is really the case.

Relevance

Thomas Kuhn argued that – in times of 'normal science' – a group of scientists share a single *Weltanschauung*, and theories will therefore be judged according to their relevance to that world-view and the presuppositions of the scientific community as a whole.

On the other hand, individual scientists, even if they are presently working on the same project, are likely to have differences in background, experience and training. It is not necessary that they all have an identical world-view and this may mean that one will accept a theory more readily than another.

Feyerabend goes a step further. He sees the process of putting forward new theories as a continuous one, and finds no absolute way of deciding between the theories put forward. But notice:

▶ The danger here is that knowledge is reduced to the prejudices or inclinations of a particular sector of the scientific community.
▶ That a theory seems relevant to the interests of one group does not guarantee that it will be relevant to others.

One can also argue, of course, that relevance is shown fundamentally in the ability to predict. Lakatos speaks of science making progress through competing research programmes, in which 'thought experiments' throw up new theories which in turn refine the questions that are being asked. A theory, for Lakatos, is examined in the light of the research programme that has developed it, and is therefore judged by its relevance to that programme.

He sees research programmes as either **progressive** (if it can lead to predictions that are then confirmed, enabling the overall coherence of the programme to be maintained) or **degenerative** (if it fails to predict, or loses its overall coherence). In other words, a research programme needs to show development – science cannot stand still.

Insight

You cannot always predict which lines of research will be successful. It is possible that an approach which now seems hopeless will one day prove useful, or the most reasonable approach in the light of present knowledge will be shown to be inadequate. Although Lakatos gives a useful way of distinguishing progressive from degenerative research programmes, it is easier to do this with hindsight. What is far from clear is how one might choose between them at the time. Science is full of surprises.

The problem is that, if theories and observations are to be judged by their relevance to an overall scheme of thought, this can degenerate into saying that theories should be judged by the extent to which they correspond to the particular insights and prejudices of the present scientific community. Add to this the theory that all observations are theory laden and that you are predisposed to see what you expect to see, and there seems no way to compare rival theories or to reach conclusions about the relative worth of research programmes. In other words, we are back to an 'anything goes' policy with respect to scientific theories.

The problem is that a situation of general anarchy when it comes to scientific claims would clearly be non-productive, since it is the desire to test out, compare theories and try to improve on them that enables science to develop.

Part of the task of the philosophy of science is to examine the way in which these issues relate to basic questions of epistemology (the theory of knowledge) and metaphysics (the fundamental structures of reality). It is always possible to ask whether a theory gives an accurate view of reality, even if, given our present state of knowledge, we are unable to answer it with any degree of certainty. Should we abandon the attempt to do this?

F. Suppe, editor of *The Structure of Scientific Theories* (University of Illinois, 1977), commenting on the changes that took place through the 1960s and early 70s, assesses the rigidity of the old positivist approaches on the one hand and the extreme flexibility of Feyerabend on the other, and concludes:

> *Good philosophy of science must come squarely into contact with the basic issues in epistemology and metaphysics; and the attempt to do epistemology or metaphysics without reference to science is dangerous at best. (p. 728)*

It is worth pausing to reflect on the implications of this last point. The rise of modern science in the seventeenth and eighteenth centuries aimed to provide a way of understanding the world that did not depend on metaphysical speculation or the authority of religious traditions. The success of its methods, based on reason and evidence, encouraged a more analytic and rational approach to all aspects of life. The excitement of engaging with science today still comes from the sense of exploring and trying to understand reality – and that, in spite of all the problems of showing whether a theory is correct or how to evaluate competing theories, promotes a sense that reason and evidence are important in all issues about what we know and how we know.

To attempt to consider the nature of reality without taking the findings of science into account invites a return to unfounded speculation and supernaturalism. Whatever else science has achieved, it has taught the value of testing out theories and remaining cautious about claims – and that is an insight that is of value to all aspects of life and philosophy.

Science and authority

Having considered the dilemma how to decide between competing theories, and whether it was possible to show that one was inherently better than another, one might well get the impression that there can be no authoritative statements in science, since everything is (and should be) open to be challenged.

This situation is made more problematic by the fact that the public generally expects science to deliver certainty. If there is a major health crisis, for example, people expect medical science to come up with an explanation and hopefully an answer. However, scientific claims generally need to be explained and qualified carefully, and there is a danger that scientists will be tempted to go along with the public's perception of their infallibility, simply in order to give an acceptable or reassuring answer. This is even more the case when there are vested interests. Both sides of the debate on global warming, the genetic modification of crops, or the safety of civil nuclear power present scientific arguments to back their case.

One might argue that many people do not want scientific explanations or advice at all. They want authoritative statements about reality; they want to know for certain whether something is or is not the case. They want to be told what they should do in order to keep well or save the planet. They want science to provide answers that are both correct and explicable in ordinary language. In other words: in turning to scientists for answers, both politicians and the general public often want the impossible.

No scientific statement can claim absolute authority. To call a statement 'scientific' simply refers to the method by which the theory on which the statement is based has been brought about. The claim to be 'scientific' is not, in itself, a claim about truth; it may be true, in the sense that it is the best available explanation for the given evidence, but that is another matter.

So how does science (in spite of Feyerabend) avoid anarchy and attempt to establish a measure of authority for its work?

PEER REVIEWING

In terms of the relationship between an individual scientist and the wider scientific community, one of the key factors in the acceptance of any theory is its publication in one of the major international journals. Once it has been made public in this way, it can be scrutinized by others working in the same field. If an experiment cannot be repeated, or the same results are not obtained elsewhere, then the theory developed on the basis of that experiment becomes immediately suspect, and being in the public domain it can be debated openly.

However, in order to get published in such a journal, an article has to be submitted to the editor who will send it out for peer review. This is the process of screening out papers that are unsuitable for publication, and it is important in maintaining the reputation of the journal in question.

The only problem with the peer review process is that it is always likely to favour those articles that conform to the present paradigm (to use Kuhn's term). There is a danger that an idea may be dismissed simply because it is too radical in its claims and methods. And, of course, if it does not get through the peer review process and achieve publication, it will not get a proper opportunity to be investigated by the scientific community. Nevertheless, whatever its limitations in this respect, little credit is given to any attempt to bypass these established steps to proper scrutiny.

Example

In 1989 two physicists, Martin Fleischmann and Stanley Pons, claimed that they had produced cold nuclear fusion. This was the attempt to create what would, if it worked, amount to a method of generating a theoretically infinite supply of energy, by harnessing the energy released in nuclear fusion. (Nuclear fusion can normally only be created artificially in very energy-intensive conditions, and in these circumstances the energy generated is far, far less than the energy used in the experiment. In nature, it is the process that keeps stars burning.)

After extensive investigations their work was severely criticized. The problem was that other people failed to get significant results from the attempts to reproduce their experiment. That did not imply that the goal of achieving cold nuclear fusion was not inherently worthwhile, simply that their attempt had not convinced the global scientific community. They were also criticized for being secretive about their research, presumably for fear that others might beat them to what was seen as a prestigious and lucrative discovery. The key to the acceptance of any theory or new discovery in the scientific community is through open scrutiny and peer review.

However 'objective' scientists try to be, and however carefully the peer group examines theories put out by those who seem to deviate from the norm, it is clear that there are times when a scientist or group of scientists feels a commitment to a particular theory or research programme, even if it not accepted by their peer group. Individual scientists are sometimes prepared to go against the general peer consensus, confident that in due course their own theory will be validated. There are many examples of scientists who have been shunned within the scientific community for their unconventional views.

Linus Pauling, who was awarded a Nobel Prize in Chemistry in 1954, and had a prestigious career, was later regarded as extremely suspect for his view that Vitamin C was a cure for many conditions, and capable of extending one's lifespan. His commitment to that view set him at odds with the scientific community. The existing achievements of an individual scientist does not therefore guarantee future acceptance of his or her views.

Sometimes, of course, a theory can be ridiculed at one time, only to be proved correct later. In 1912 Alfred Wegener (1880–1930) was unsuccessful in arguing for his theory of continental drift, and when Newton proposed his theory of universal gravitation, it was hugely controversial because neither he nor anyone else at that time could understand how gravity could operate between bodies without some physical medium joining them. It was only because his theories were successful in their predictions that it was assumed that they were correct and they gained acceptance. So even the most prestigious of theories may have difficulty being accepted initially, simply because of their radical nature.

The scientific community tends to be quite conservative. Only when there is cumulative evidence of inadequacy is a paradigm or research programme finally abandoned and replaced by another. To some extent, this is justified on any pragmatic or functional view of scientific theories – if a set of theories, or paradigm, is working well and yielding results, there is a tendency to keep to that paradigm until it finally gives way to another which clearly offers greater explanatory power.

Insight

To the innovative thinker, this may smack of authoritarian conservatism. To the scientific community as a whole, however, it remains a safeguard against the rash abandonment of useful and established theories in favour of every novel suggestion.

KEEP IN MIND...

1 The key question is how we decide between competing scientific theories.

2 Our observations are influenced by our existing ideas.

3 Theories are underdetermined if there is insufficient evidence to decide between them.

4 Theories compete with one another within and between research programmes.

5 Factors other than truth determine which theories succeed.

6 Instrumentalism judges theories by their ability to predict, and thus their usefulness.

7 It may be worth sticking with a useful theory, even if we know that it is wrong.

8 Theories may be assessed in terms of their relevance to the background assumptions of the scientific community.

9 Research programmes can be progressive or degenerative.

10 Theories are normally submitted for peer review and published in order to be assessed by the wider scientific community.

7

....

Predictability and determinism

In this chapter you will:
- *consider whether the future is absolutely determined by the past*
- *explore probability theory*
- *look at issues raised by complexity and chaos theory.*

Is the world chaotic and unpredictable, or is it ordered according to physical laws that we can understand on the basis of experimental evidence? Does everything happen by chance, or is it determined by causes, only some of which we may know? Is freedom and chance an illusion generated by our ignorance of all the facts?

Presented starkly, such questions have implications for religion and morality as well as for science and for our general understanding of the nature of reality (metaphysics). We seem to be offered a straight choice between freedom and determinism. However, matters are not that simple.

First of all, we have seen the way in which scientific laws are justified by a process of induction, based on experimental evidence and observed facts. They may therefore be regarded as the best available interpretation of the evidence, but not as the only possible interpretation of the evidence. They do not have the absolute certainty of a logical argument, but only a degree of probability, proportional to the evidence upon which they are based.

Having said that, when we consider the way in which science developed in the seventeenth and eighteenth centuries, it is clear that reason, order and predictability were at its heart. To interpret events as a matter of luck, fate or the capricious will of the gods was seen as superstition, to be set aside by the reasonable explanations of the emergent sciences.

In this chapter we shall consider the view that the world is predictable and, taken to its logical conclusion, that everything is determined. We shall then look at the inherent weaknesses of that view, in terms of scientific method and also in terms of those features of the world known to science which cannot be fitted into such a view.

We shall also need to look at the whole issue of whether it is possible to deduce laws from statistical evidence, and how such general statistical laws relate to the causal or free actions of the individuals that go to make up such statistics. This, of course, is particularly appropriate in considering the social sciences.

Predictability and determinism

With the development of modern science came a general assumption that the world was predictable and capable of being understood rationally. This was, of course, fundamental to that whole movement in thought that we tend to refer to as the Age of Reason. As we saw in the historical introduction, the development of the physical sciences, leading up to the work of Newton, presented the world as essentially a mechanism, whose workings were understood in terms of the laws of nature.

Most philosophers agreed that everything was theoretically predictable. Thus, David Hume thought of an apparent chance event as merely a sign that we were unable to know all the forces operating upon it.

> *'tis commonly allowed by philosophers that what the vulgar call chance is nothing but a secret and conceal'd cause.*
>
> David Hume, *A Treatise of Human Nature* (1739–40)

Kant equally affirmed the necessity of seeing everything in the world as being causally determined, since it was something that our minds imposed upon experience. For the purposes of this argument, it is less important whether those regularities are empirically verified or imposed by the mind; the essential thing is that there becomes no place for the chance occurrence or random event.

The classic expression of this view is given by **Pierre Laplace** (1749–1827), who held that, if one were to know all the causes operating, then, having known one single event or thing, it would be possible to demonstrate both everything that had taken place and everything that would take place. The whole universe comprised a single, predictable mechanism:

> *Given for one instant an intelligence which could comprehend all the forces by which nature is animated and the respective situation of the beings who compose it – an intelligence sufficiently vast to submit these data to analysis – it would embrace in the same formula the movements of the greatest bodies of the universe and those of the lightest atom; for it, nothing would be uncertain and the future, as the past, would be present to its eyes.*
>
> A Philosophical Essay on Probabilities (1816)

Now there are two senses in which one can speak of determinism:

1 the theoretical ability to account for every human choice
2 the view that everything is part of a chain, or web, of causation.

While the first is relevant for morality and questions of human freedom, the second is of principal interest for the philosophy of science. But let us examine the first for a moment, since it impinges on our appreciation of the second:

▶ Is your apparently free choice actually determined by the past?
▶ Does your apparently free choice determine the future?

Choosing to turn right precludes all the events and experiences that might have taken place if one had chosen to turn left. From the perspective of the agent of choice, the future is affected by that choice only from the moment it is made, until then its possibilities remain open. However, making a choice is a matter of weighing possibilities, values, inclinations, early training, experience, hopes, fears, whatever. All of those things form the background to the apparently free choice, and if any of them were different, the choice you make might also be different. The question is, do they determine that choice, to the extent that no other choice is possible given these circumstances? Does our experience of freedom reflect actual freedom?

If we combine this with the second sense of determinism – that everything that happens is totally predictable in terms of its antecedent causes – then our experienced freedom is either an illusion, or it belongs to some order other than the physical (in other words, that mental operations are not susceptible to physical laws). This, as we shall see later, is challenged by neuroscience.

Insight

Morality and personal responsibility are based on the experience of freedom. They are not automatically negated by the claim that all physical operations are causally determined. Even if my choice is predictable, it remains my choice.

Absolute determinism works on the assumption that it is possible, at least in theory, to give a full causal account of everything. However, if we adopt the approach recommended by Karl Popper – using conjecture and always allowing for falsification – we can never claim to know a full causal series. At most, we present an explanation as the best available, never as the ultimate or final one.

SOME HISTORICAL PERSPECTIVES

Democritus, a Greek philosopher of the fifth century BCE, was an atomist (see Chapter 2). He considered everything to be comprised of atoms, which could not have emerged out of nothing, and which he therefore considered to be eternal. However, observing that everything in this world is subject to change and decay, he concluded that physical bodies were but the temporary gathering together of these eternal atoms, and that they changed as their component atoms dispersed to form other things. He therefore considered it theoretically possible to predict how each and every thing would behave, simply because that behaviour was determined by the atoms of which it was composed.

This view was taken up by the Epicureans who saw the whole universe as a single determined mechanism, operating on impersonal laws. For them, it had important implications for human morality and self-understanding: if we are but temporary, composite creatures, with our life determined by these impersonal laws, it makes no sense to worry ourselves about some ultimate meaning or purpose. We are free to decide how we should live, and it is therefore perfectly valid to decide that the chief aim in life is to be the quest for happiness and wellbeing.

From the same period, you have a very different approach taken by the Stoics. They held that everything in the universe was ordered by a fundamental Logos ('reason' or 'word'), that one simply had to accept what could not be changed, and that – in those things that involved personal choice – one should seek to align one's will to the Logos, thus fulfilling one's essential nature.

Insight

This period of Greek philosophy thus presents two very different views: one that the world is impersonal and constantly changing, the other that it is ordered by reason. The Epicurean view points to determinism; the Stoic to the view that, within a world where many things are beyond our control, we can nevertheless effect a measure of change – the skill being to know what we can change and what we can't.

The mechanistic view provided by Newtonian physics suggested that everything may be described in terms of fundamental 'laws of nature' which operate with mathematical precision. The clear implication of this is that, if all the laws were known, it would be possible to predict exactly what would happen. If something unpredictable happens, it simply indicates that there is some new law of nature, of which we are presently unaware, operating in this situation. Thus, science came to assume that everything had a cause – or indeed a large number of causes working together – which determined exactly what it was, even if that determination could not be exhaustively demonstrated.

How then do we account for the experience of human freedom? What does it mean to choose in a world where everything is determined? Central to the way in which science coped with such questions was the view, put forward by Descartes, of a radical mind/body dualism. The body was extended in space and time and was controlled by the laws of nature. The mind, although linked to the body (through the pineal gland), was not extended in space, and was therefore free from the determination of physical laws. It was this Cartesian dualism, along with the rise of modern science, that resulted in the perception of a mechanical universe, totally conditioned and determined, every movement theoretically predictable. Human beings could contemplate and act within such a world with apparent freedom, since their minds were not part of it, the mental realm being quite separate from the physical.

The philosopher and mathematician **Gottfried Leibniz** (1646–1716) presents what amounts to a rather different angle on determinism. He argued that a change in any one individual thing in the world would require that everything else be changed as well. It might be possible for the whole world to be different, but not for the rest of the world to remain as it is if this one thing were changed. Why then do we experience freedom? His answer is that, not having an infinite mind, we cannot see the way everything works together. We therefore cannot actually know all the factors that control our actions, and therefore are able to believe that we are free.

Insight

Does this sound familiar? Leibniz seems to have taken the 'Duhem–Quine approach' (see Chapter 4) two hundred years before either Duhem or Quine!

As we have already seen, Kant distinguished between:

▶ **phenomena** (things as we experience them) *and*
▶ **noumena** (things as they are in themselves).

He argued that the mind understands phenomena by means of the concepts (space, time and causality) that it imposes upon experience. Thus, we can say that everything has a cause, not because we have been able to check absolutely everything, but because that is the only way our minds can understand the world.

With Kant, choices are determined by our desires, beliefs and motives. Once made, they have inevitable consequences. Thus, from the standpoint of the choosing subject there is freedom, but from the standpoint of the observer (both in terms of observation of motives, and in seeing the way in which actions dovetail into chains of causes) everything fits into a pattern in the phenomenal world, a pattern which inevitably suggests causal determination.

Insight

Notice that both Leibniz and Kant are writing against a background of a rational and scientific view of the world (as expressed by Newtonian physics) and struggle to account for the experience of freedom in such a world. For Kant, it is made possible only by making a fundamental distinction between things in themselves and things as we perceive them: the former free, the latter determined. For Leibniz, freedom is an illusion born of our ignorance of the totality of causes acting upon us. Whereas for Descartes and Kant, freedom is real, for Leibniz is it not – and that makes his view rather closer to a more radical view of determinism that developed through the nineteenth and twentieth centuries.

In *The Riddle of the Universe* (1899), Ernst Haeckel argued that everything, including thought, was the product of the material world and was absolutely controlled and determined by its laws. Freedom was an illusion; scientific materialism was the only valid interpretation of reality. This view reflected both the success of science by the end of the nineteenth century, and also its limitations. By the end of the twentieth most scientists were being far more cautious in their claims.

In looking at twentieth-century science, we have already seen that some things in physics (e.g. the behaviour of subatomic particles) or in biology (e.g. genetic mutation) seem to happen in a random way. Chance replaces strict predictability. On the other hand, once a chance event has taken place, other things follow from it in a necessary fashion. A classic exposition of this dual shaping of reality was given by Jacques Monod (1910–76) in his book *Chance and Necessity* (1972). Monod claims that the whole edifice of evolution, which to some appears to be the product of design, can be accounted for entirely by the operation of physical laws upon the many chances that are thrown up by genetic mutation.

My experience of freedom may thus be explained in terms of chance (the particular circumstances in which I experienced myself as making a free choice) and necessity (the factors that, with hindsight, I can see as determining my choice).

The uncertainty principle

Heisenberg's uncertainly principle – that it is possible to know either the position or the momentum of an particle, but not to know both accurately at the same time – is often cited as an example of freedom at the very heart of modern quantum physics, and therefore that determinism is generally inadequate. It is unwise to press this argument, however. We know that nature can be regular and predictable on the larger scale, even if at the subatomic level individual particles are undetermined. In other words, there is no contradiction in arguing that, at the level at which quantum theory comes into play, reality may be non-deterministic, while classical physics remains deterministic.

Probability

We saw, in looking at the classic formulation of the problem of induction, that there can be no such thing as absolute certainty, but only increasingly high degrees of probability. To formulate a general theory from observations requires a leap beyond the evidence itself, based on the assumption that the universe is predictable and uniform, and that the more evidence is gathered for something, the more likely it will be that further evidence will confirm it. But no finite number of instances can logically require us to conclude that all evidence will confirm a theory.

Apart from Monod, all the thinkers mentioned so far in this chapter lived prior to the end of the nineteenth century. Issues of freedom and determinism –particularly as they have come to be studied in ethics and the philosophy of religion – still rely significantly on arguments from that period. However, during the nineteenth century there were significant changes – particularly in the gathering and analysis of information about populations – that led to a very different approach to scientific theories and determinism, an approach which allowed indeterminism and probability to operate at an individual level, while claiming overall statistical theories to apply to large numbers.

During the nineteenth century there was a considerable increase in the recording of statistics about human life and death. Using these, it became possible to formulate theories about human behaviour, in the same way as experimental evidence had provided the basis for the physical 'laws of nature'. As we shall see later, this led to the establishment of the social sciences, as humankind became increasingly the object of scientific examination.

Ian Hacking, in *The Taming of Chance* (1990), has shown this increasing use of statistical data and the way in which what, on an individual level, appeared to be a simple matter of random chance, became a piece of data to be interpreted as part of an overall statistical law. This had profound implications for an understanding of determinism, for we now have laws which are not based on an analysis of individual events, but on statistics which summarize large numbers of events. And, of course, statistics cannot be used to prove what happened in an individual situation, only to show its statistical probability.

Example

In a general election, after the first few results are in (or even after an 'exit poll' has been taken), commentators start predicting the result. In seat after seat, patterns are watched and tabulated. It is deemed likely that the people voting in any one polling station will behave just like others up and down the country. However, for each individual entering the booth, there is an absolutely free choice.

Our key question therefore: How then can such freedom be reconciled with statistical probability?

The sociologist Durkheim said of this phenomenon:

> *Collective tendencies have an existence of their own; they are forces as real as cosmic forces*
>
> Émile Durkheim, *Suicide* (1897)

Clearly, this has implications for looking at the human sciences, but notice the broader implication – that scientific laws can operate in terms of statistical averages, rather than on the ability to predict what will happen on each and every occasion.

For Durkheim, the 'laws' that could be known statistically were rather like a force of a certain strength sufficient to produce a result in a limited number of cases, but not overwhelming of the individual. Some respond to that force, others do not – but statistically the number responding is predictable.

Insight

Statistical analysis can show 'law-like' tendencies. There is a tendency for trees to be blown over in a severe storm, but only some will actually fall. The blowing of the wind is not the only factor in the fall of an individual tree, but it will increase the probability of it.

In effect, this approach amounts to a theory of influence. Statistical laws are in fact descriptions of degrees of influence that apply to individuals within their field. The degree to which this influence is effective will depend upon many other factors. Thus the individual is not determined or predictable, while the overall trend is quantifiable.

Modern political and social theory gives many examples of this sort of argument. Young people from deprived backgrounds may be deemed more likely than others to commit acts of vandalism. That does not explain the individual choice to commit such acts; it simply suggests that deprived backgrounds tend to exert an influence in that direction.

Hence, statistical probability gives us a way of assessing the likelihood of a piece of evidence. The inductive method had an element of probability built into it, since it was never able to find all the possible evidence for a theory, but – based on the evidence available to date – it was possible to assess the probability that the next piece of evidence would conform to it. In our discussion of the issues of how theories are assessed against one another, falsified or replaced, it is clear that no theory is established with the degree of certainty that places it above the possibility of challenge.

Hume, in claiming that a wise person proportions belief to evidence, presents the case for dealing with degrees of probability, so dealing with the issue of probability is not new. What is new, from the nineteenth century, is this use of statistical probability, where the statistic itself becomes the evidence. Hence it is not necessary that each and every piece of evidence is required to conform to a theory for that theory to be confirmed (not even Popper, in the strictest interpretation of how falsification works, would claim that), but that, taken statistically, the theory should reflect the balance of probability in the evidence.

Names

Your name is most improbable. Whatever you are called, there are a very large number of names that you were not given – so the chance of you having your name is very small indeed. On the other hand, given the wishes of your parents and perhaps other relatives, favoured people after whom you might be named, family traditions, names with special meaning within your culture or religion and many other factors, the probability of your particular name being chosen increases. You age may also be an indicator, since fashions in names come and go. Indeed, the very fact that you were born either male or female cut by almost half the number of choices open to your parents!

When every factor is weighed, including those coincidences for which no explanation can be given, your naming might appear easily explicable, perhaps inevitable. A stranger's name, on the other hand, may appear to me to have been chosen randomly, for I have no way of assessing its probability.

Much the same applies to physical phenomena. On the one hand, they may seem infinitely unlikely; on the other, under scientific investigation, they become inevitable.

EQUALLY IMPROBABLE!

If you claim that a law of nature – for example, one of Newton's Laws of Motion – has universal application, then it must apply to an infinite number of situations. Every time you test one of these, the law is confirmed and therefore becomes a little more probable. However, even if it can be shown to apply to an almost infinite number of such situations, there will always be an infinite number that have not been tested (for infinity, less a finite number, is still infinity). You might therefore argue that all universal theories are equally improbable and almost equally underdetermined!

Such a dilemma makes it important to find a method of assessing the probability of something being the case, given the evidence to hand – and it is to this which we now turn.

CALCULATING PROBABILITY

Thomas Bayes (1702–61) argued that the probability of something happening could be expressed mathematically, and that the resulting probability equations are needed in order for inductive reasoning to work. '**Baysianism**', the general theory of probability that became influential in the philosophy of science in the twentieth century, is based on his work.

Basically, a rational person believes X to be the case to the degree that he or she believes 'not X' not to be the case – and thus the sum of the probability of these two opposites always equals 1. In a sense, this is obvious. But Baysian approaches take it a further step and assess what evidence it takes to change that belief. In effect, a single piece of evidence Y should not automatically refute all belief in X. Rather the belief that X is the case should have a probability proportionate to the acceptance of both X and Y. You do not simply scrap and rethink all your beliefs with every new piece of evidence produced, but moderate them on a scale, taking that new evidence into account.

> **Insight**
>
> If I have a theory X that predicts 'Y will happen', the degree to which the evidence for Y confirms the theory X will depend on the likelihood that Y would have happened anyway, even if the theory X were wrong.
>
> If the latter is the case, the theory X is not well confirmed by that evidence, even if it predicted it. But if Y is extremely unlikely to happen unless X is correct, it offers very strong confirmation.

To put it crudely – it is no use expecting your new theory to be confirmed on the basis of its prediction that the sun will rise tomorrow! If it is going to happen anyway, it cannot significantly increase the probability of your theory being correct.

Let us return to the familiar example of Eddington's eclipse observations, which confirmed Einstein's theory that light coming from distant stars would be bent by the gravitational pull of the sun. Without Einstein's theory, the probability of the light from the stars being bent in this way was very low indeed, since light had always been believed to travel in straight lines. Therefore the observation that this was in fact the case provided the strongest possible confirmation that Einstein's theory was correct.

The most popular version of Baysianism is the subjective Baysian approach. This argues that the probability calculations refer to the degrees of belief that scientists have in a particular theory. In other words, it provides a way of ascertaining whether it is reasonable to believe a theory to be true, not whether it is actually true.

Chaos and complexity

When considering probability, you look for patterns in large numbers and draw out statistical conclusions, from which you can make predictions. In chaos theory that situation is in some ways reversed, for it explores the way in which a minute change can bring about radically different effects, making prediction impossible.

It was given popular exposure through the work of Edward Lorenz (1917–2008) in the 1960s, who examined the effects of turbulence in dynamic systems, like weather forecasting. He described chaos as a 'sensitive dependence' on initial conditions, and produced the most popular image of this: that the flapping of a butterfly's wing in Brazil might cause a tornado in Texas. In effect, this theory argues that chaos results from a system of regular feedback in which an initial variation is magnified again and again with unpredictable consequences.

An example: ten-pin bowling

However skilled you may be at bowling, there will always be minute changes in the angle at which you release the ball that will be magnified as the ball travels the length of the lane. As it strikes, the first skittle falls back either slightly to the right or the left, and the ball is deflected slightly in the other direction. From then on, within a fraction of a second, skittles start falling in different directions, sometimes hitting others as they fall. The differences in the final arrangement of skittles each time is difficult to predict from the slight variation of angle as the ball leaves the bowler's hand. Even those who can achieve strike after strike actually achieve a different strike every time, for the skittles will never fall in exactly the same way twice.

In a truly sensitive system, the amount of variation goes far beyond that of the crude bowling analogy. Every tiny change is magnified again so that, even though each of those changes is mathematically simple, prediction becomes impossible. Hence: chaos.

However, small changes can sometimes work together in quite a different way. Complexity Theory is the study of structure, order and stability and is particularly associated with the work of Ilya Prigogine (1917–2003) (*Exploring Complexity* (1989)), and it has important implications for our understanding of many areas, including evolution.

Complexity theory explores the way in which small changes gradually accumulate in such a way that a pattern emerges. It can now be shown, for example by computer simulations, how elaborate life forms can assemble themselves by the operation of simple choices. This would go to explain how nature can have the appearance of design without the need for an external designer. The complex patterning, which we see as design, is the product of a large number of simple operations.

We also need to recognize that the principles that determine the operation of a complex entity only emerge at that degree of complexity, not at a lower level. In other words, you will never discover anything about human behaviour from the analysis of a small piece of human tissue. A study of complexity is the very opposite of a reductionist approach. A complex entity works at the level of its maximum not its minimum complexity. If you want to know about patterns of congestion on a highway, don't start by analysing the internal combustion engine. Look at the way cars are used.

COMPLEXITY AND DISORDER

There might appear to be a problem with the idea that things can gradually increase in complexity and take on the appearance of design, namely that this appears to be at odds with the Second Law of Thermodynamics, which shows that everything is gradually losing energy and running down, with a consequent increase in randomness and disorder.

The answer to this difficulty lies in the difference between open and closed systems. An open system is one that is able to take in energy from its surroundings, and it is this energy that enables its complexity to be maintained. In a closed system there is no way of replenishing the energy lost and the result is increased disorder.

Example

If you were sealed off from your environment, you would die and your body would gradually disorganize or decompose. It is only because you receive nourishment and oxygen that you can maintain yourself as a complex being. Therefore, even though the universe might gradually wind down, with increasing disorder and entropy, within it you can have pockets of increasing complexity, fuelled by energy from their surroundings.

KEEP IN MIND...

1 Hume considered that ignorance of causes gave the appearance of chance.

2 Determinism is the view that everything is, in theory, predictable and inevitable.

3 Determinism may be a feature of the real world, or of our way of understanding the world.

4 Heisenberg's uncertainty principle operates only at the subatomic level; larger entities remain predictable.

5 Statistics show general trends without thereby predicting individual choices.

6 Thomas Bayes devised a method of calculating probability.

7 Chaos theory argues that small changes may initiate sequences whose results are impossible to calculate.

8 Complex patters may be built up over a period of time from small individual changes.

9 Complexity is natural, although it gives the impression that it is the product of conscious design.

10 Increasing complexity is a feature of open systems.

8

The philosophy of biology

In this chapter you will:
- *examine the theory of natural selection*
- *explore philosophical issues raised by evolution*
- *consider the impact of genetics.*

Every branch of science raises particular philosophical issues and biology is no exception. The scientific case for evolution by natural selection was controversial, both because it rendered unnecessary the idea of a designer god and also because it showed an essential link between humankind and other species, thereby denying to *Homo sapiens* a unique place in the scheme of things. The idea that it is right or necessary to infer the existence of an external designer from the appearance of design was criticized by David Hume in the eighteenth century, so the matter was controversial long before Darwin published his theory of natural selection.

The issues raised by natural selection are brought into closer focus in considering genetics, which, quite apart from showing the degree to which all species are related to one another, has applications which raise ethical dilemmas. In particular, the ability of genetics to offer a way to modify living forms has enormous implications for the control that humankind has over the environment as well as over its own health and the development of the human species.

Natural selection

In terms of its origins, Darwin's theory of natural selection seems (as many brilliant new insights do with the benefit of hindsight) to be quite straightforward. He started his argument by observing the amazing variety of, for example, dogs or flowers that humans were able to bring about through selective breeding. He then asked whether a similar process might account for the wide variety of living forms in nature. What he needed was to find a mechanism within nature as a whole which corresponded to the selective breeding carried out in the domestic sphere by the gardener or farmer – a 'natural selection' to parallel their 'artificial selection'.

The process of struggle for existence (as it had been outlined by Malthus) in the face of limited resources led to his perception that – in any such competitive situation – certain favoured individuals are able to survive and breed, at the expense of those not so favoured. He saw also that the key feature of success was the ability of a species to adapt to its environment. Those variations in individual members of a species that favoured them in competing for limited resources were likely to enhance their chances of survival into adulthood and therefore of successful breeding. The result of that would be the wider dissemination of those particular variations in the population as a whole. Thus, the process of selective breeding by humans was being carried out by large, impersonal forces within the natural world.

Initially, Darwin's theory met opposition not primarily on the grounds that there was anything wrong with his argument, but in the light of the more fundamental question of whether any theory for the development of species was necessary at all. Largely on religious grounds, but also as a common-sense response to the observation of the amazing variety of living forms, it was assumed that each species had been created distinct from the others. Much of science had been involved with the listing and classification of species. Rather than developing in such a way as to adapt to their environment, it was assumed that each species was designed in order to make its continued existence possible.

Darwin's argument provided a mechanism that rendered all reference to external design irrelevant. The mechanical and mathematical approach, which had swept aside the older medieval view of a purpose-full world, now seemed to engulf all living things, humankind included. That required a major rethinking of previous philosophy, and initial opposition to Darwin was not limited to those of a conservative religious disposition.

The other problem with Darwin's theory was the lack of fossil evidence for transitional states between species. Of course, there is a natural explanation for this lack of evidence. Before humans developed burial rituals and the embalming of corpses, most species simply died and were left to rot, or were eaten by other species. Left to its own devices a body is soon reduced to bones, and those bones eventually become dust, so that no trace remains. It is only in the rare circumstances of a body being preserved (e.g. by being trapped in a bog) that we find remains.

One might also argue that evolution did not take place incrementally, but in short bursts, interspersed with longer periods of relative stability. Thus, given that the transition between species is faster than the persistence of that species once formed, and given that, of the millions of specimens that live, only a handful are going to end up as fossils, it is only realistic that we should find few, if any, fossils that show a transitional state between species.

Darwin's theory is presented in a logical way, and in effect it invites the reader to go along with it, and then examine the balance of probability that it is correct. Natural selection was not born out of the interpretation of fossil evidence, but out of observation and analogy which was then supported by evidence.

Subsequent evidence

In *The Beak of the Finch* (1994), Jonathan Weiner describes a 20-year study of finches on one of the Galapagos Islands, the place where Darwin first gathered the information that led to his theory of natural selection. Weiner gave detailed information about the small differences in the length of beaks, showing for example that in times of drought only those finches with the longest beaks could succeed in getting the toughest seeds, and therefore survive to breed. What is amazing is that only a very small difference (less than a millimetre in length) in beak size can make the difference between life and death in times of severe competition for limited food. DNA samples taken from the blood of various finches showed the genetic differences that corresponded to their physical abilities and characteristics of each type of finch. Here we have a modern piece of research which provides additional data to illustrate Darwin's original observation.

However, that data comes to hand simply because the modern scientist already knows what to observe and record. It is a good example of using a theory in order to select relevant data. That is not, in itself, an illogical thing to do – indeed, it is the only thing one can do.

Quite apart from detailed studies like Weiner's, there are many examples today of the process of natural selection. Whereas the evolution of different species happens over long periods and is therefore not open to straightforward observation, there are some forms of life that mutate and establish themselves very quickly. Every now and then a new strain of 'flu' emerges and is able to flourish because it is sufficiently different from the previous one to render existing immunization programmes ineffective. Faced with a hostile (well-immunized) environment, only those varieties of a virus that are not affected by immunization will survive and develop.

A TAUTOLOGY?

There is a common misunderstanding of Darwin's theory of natural selection, related particularly to the phrase 'survival of the fittest' (a phrase which, incidentally, was coined by Herbert Spencer rather than Darwin). This takes the form of the criticism that 'survival of the fittest' is a tautology; in other words, that the phrase is a proposition defining what one means by fitness.

Now if this were true, it would be a significant criticism, but it is based on a misunderstanding. The expression 'survival of the fittest' can indeed describe Darwin's theory, but it is not a proposition (if it were, it would indeed be a tautology). In other words, he does not argue that the fittest 'should' survive, nor does it ascribe a quality 'fitness' to those who survive. The phrase 'survival of the fittest' is simply a summary of an observed process.

In other words, within a population that includes characteristics that vary, those variations that happen to enable an individual to survive to adulthood, will enhance the chances that the individual will breed, and will therefore increase the incidence of those variations in the next generation.

A related question was raised by Karl Popper. He pointed out that, if a species did not adapt to its environment, it would not survive; hence every species that continues to exist must be fit in the sense that it must have been able to adapt. That being the case, everything is compatible with the idea of 'survival of the fittest' since it is impossible to show that something has survived without adapting. And, of course, Popper regards it as essential that, for a theory to be genuinely scientific, it must be capable of falsification. However, in response to Popper's point, one can argue again that 'fitness' is simply a way of describing the ability of a species to adapt and therefore to survive.

Insight

The point of natural selection is not that those who have beneficial characteristics have a better chance of survival – that is simply an observation – but that it is a process capable of producing long-term changes in a species.

Causes and purposes

One of the interesting offshoots of discussion of natural selection is a comparison of the evolutionary and purposive aspects of biological phenomena. As we saw above, one of the features of natural selection is that it provides an impersonal mechanism for producing the appearance of design. But how does that account for those features of a species which appear to have a particular purpose?

Some animals, for example, have colouring which helps them to blend into their surroundings, thus enabling them to hide from predators. In common speech, one might say that the 'purpose' of that colour is for defence. Similarly, I need to keep my body temperature within certain limits. If I get too hot, I go red, the pores of my skin are dilated, and I sweat, thus releasing heat. From this perspective, each feature of a living thing has a purpose, which is its contribution to the overall aim of maintaining the life of the individual animal or plant.

Natural selection appears to interpret such features in terms of cause rather than purpose. In other words, I have certain qualities, and bits of me perform certain functions in order to promote my wellbeing, simply because – through the process of natural selection – they have been qualities and functions that have promoted survival, and have therefore survived and spread within the gene stock of the species.

Example

In the green of the jungle, vulnerable animals that are brightly coloured are quickly seen and eaten. Those that are green, tend to get overlooked, grow to adulthood and breed. Hence the colour whose 'purpose' appears now as camouflage is naturally selected.

There remain some fundamental philosophical questions that can be asked of evolution. Given the way in which species change, can evolution be said to be progressive? Does it have some final goal and, if so, how can we determine what it is?

The philosopher and Jesuit priest **Teilhard de Chardin** (1881–1955) argued that evolution produced successively more complex life forms. He suggested that evolution could be thought of as a cone with atoms at the base, rising through molecules, cells, living creatures, up to humankind, and then on to a theoretical point of total convergence, which he termed Omega. Although his argument contained elements that cannot be doubted (that evolution has, in fact, produced life forms with increased complexity), his vision is one that is inspired and shaped by a teleological (end-orientated) and religious purpose.

This directional scheme for evolution cannot be justified by such evidence as would generally be regarded as acceptable within the scientific community, however much his vision might prove inspirational for those seeking a sense of overall direction and purpose. In general, although natural selection produces members of a species that are increasingly well-adapted to their environment, it is difficult to use this as a basis for giving evolution some overall direction, not least because environments differ so much.

Insight

The actual 'progress' of individual species from earlier forms may be assessed retrospectively in the light of where they are now. It is quite another matter to speculate on future directions each might take, not least because natural selection operates in the context of an ever-changing and competitive environment.

Environments are complex things. Which is more important – to adapt to lack or excess of water, or to have a colour that blends into the background and thereby makes one less vulnerable to predators? Any one species may be wonderfully adapted to one aspect but not another. Survival may require multitasking in the art of adaptation.

The genetic basis of life

The theory of natural selection showed how, given a number of variations within individuals in a species, those with particular advantages would survive and breed. What Darwin did not know was the reason for the small physical changes that had such a profound effect on the fate of individuals, and through them on evolutionary progress as a whole. Genetics subsequently revealed the way in which those random mutations occur.

Deoxyribonucleic Acid (DNA)

All living things are composed of chemical substances, of which the nucleic acids (RNA and DNA) and proteins determine the way in which living cells are put together. In the nucleus of each cell (except for sperm, egg and red blood cells) there are 23 pairs of chromosomes, which contain the DNA, made up of two strands of chemical units called nucleotides, spiralled into a double helix. The 'information' given by the DNA takes the form of sequences of nucleotides (genes).

Human DNA comprises about 30,000 active genes, which give the code for making 20 different amino acids, which themselves produce the proteins, which make us what we are. But these are only a small part of the total information contained in DNA. Between 95 and 97 per cent of the nucleotide sequence is regarded as 'junk DNA' and is not used in the construction of a human being. Most has been left behind in the evolutionary process, giving theoretical instructions that are no longer needed and are therefore deleted.

The strands of DNA are around 6 foot in length, and the genetic information is given by sequences of the four chemicals (bases) – adenine, thymine, guanine and cytosine. The whole human genome is thus a digital code made from those sequences of chemicals – a total sequence of around 3.5 billion pieces of information.

As cells reproduce, the chromosomes work in pairs to copy the information required. Generally, where one of them produces a defective copy, the other provides a good one. From time to time, however, a defective copy survives, and (in response to defective instructions) one of the amino acids in a protein chain is changed for another. This is termed a mutation. Mostly, these lead to irregular growths in tissues, but if a mutation takes place in a germ cell (which is rare) then that change can be inherited.

Mostly surviving mutations either make no difference, or are neutral, as far as the individual possessing them is concerned. Just once in a while they make a positive difference, work well alongside the rest of the genes in the 'gene pool' of that particular species and thus improve that individual's survival chances. Thus, genetics provides a crucial underpinning for natural selection.

THE INTERCONNECTEDNESS OF LIFE

One of the major events in the first years of the twenty-first century was the analysis of the human genome, a public draft of which was published in June 2000. The intention behind that project was as much practical as based on 'pure' science – namely, that a knowledge of the operation of each of the human genes could lead to the prevention of inherited diseases, such as cystic fibrosis. But its publication raised again some fundamental questions about the relationship between humankind and other species.

One of the most interesting things about the human genome is the recognition of just how many genes we share with other creatures, even very simple ones, and the way in which the same gene can perform similar functions in different species. This reinforces the general view of evolutionists that all forms of life are closely linked.

Insight

If we seek evidence that species are linked, we can look at the genes they have in common. Chimpanzees have DNA which is 98.9 per cent identical to that of humans.

Identical genes play similar roles in different species, which suggests that the species can be traced back to a common ancestor, from whom each has inherited genetic material. This means that we have a new route by which to trace the development and splitting off of species from one another, since the degree to which two species share common genetic information is a good indicator of how long ago they diverged from one another.

The overall view of humankind and its place within the biological world has been profoundly changed by the recognition of our shared genetic heritage. This has implications that are more appropriately explored in terms of the philosophy of religion or ethics.

Some implications of genetics

In terms of our self-understanding, it is difficult to imagine any single scientific discovery equal to the setting down of the human genome. The genome represents the set of information required to create a living human being. That life can be the product of a sequence of bits of information is remarkable in itself – it defines our uniqueness (we are all different) and interconnectedness (all living things share common DNA).

Basically, genetic information determines the form and operation of living things – and thus must play a significant part in their medical history. It may indicate a predisposition to certain forms of illness, or predict genetic abnormalities in the unborn. It therefore offers a degree of control over the human body that is more fundamental and subtle than the traditionally invasive operations of medicine. The detection and replacement of faulty genes allows the body to learn to heal itself.

Insight

Genetics touches on the sense of identity of each individual. On the one hand, forensic examination uses the fact that genetic information is unique to each individual to identify him or her from sample tissues. On the other, we know that most of our genes are shared with all other living things. We are simultaneously unique and interconnected.

One of the key questions related to genetics is the degree to which we are the product of genetic information, as opposed to environment or training. This, of course, has social and political implications, especially for those who take a basically socialist (or Marxist) view that human life is largely determined by material and social conditions, and therefore that people can change themselves by changing their circumstances. The fear expressed from this point of view is that the more human behaviour is conditioned by genetic factors, the less one can be held responsible for it. This, of course, is yet another form of the freedom versus determinism debate.

One of the great turning points for the scope of philosophical enquiry came in the 1970s, when there was a progressive recognition that philosophy needed to address a number of social and political areas of life, rather than (as had been the case for the decades before that) being focused more narrowly on the meaning and use of terms. In other words, philosophy was expected to have something to say about human situations, not merely about the language other people used when commenting on those situations. This movement was given particular impetus in the fields of medical and nursing ethics – and these are, naturally enough, still the areas in which the implications of genetic research are felt most acutely.

Part of the anxiety expressed about treatment at the genetic level has to do with the pace at which the knowledge is being gained and applied, and perhaps the fear of a 'Frankenstein' approach to humanity, where science seeks to construct life artificially. This enlarges the scope of the existing debate about the moral legitimacy of in vitro fertilization (IVF), for example, where conception becomes possible in circumstances where a couple would formerly have remained childless. Should there be a limit to the way in which technology is used to change medical conditions or bring about a result that would not have been possible in a natural environment? Of course, those who would place limits on treatment using genetic manipulation need to specify at what point they would wish medical technology to be limited, since all medical intervention is aimed at frustrating the natural course of illness.

Key questions

▶ Do I control my genes, or am I controlled by them?
▶ What are the implications of the predictability of future health? Should genetic information be provided to insurance companies, or be required as a condition of employment?
▶ What are the implications of a genetic revolution in the diagnosis and treatment of illness?
▶ What implications, if any, does genetics have for the sense of human freedom?

SOCIOBIOLOGY

Sociobiology, published in 1975, is a book that explores the biological and evolutionary factors in the development of society. Its author, E. O. Wilson (1929–), analysed social behaviour from ants to humans. Had he limited his work to animal behaviour, he would have avoided criticism, but his last 30 pages were devoted to looking at how his theory could be applied to human society. It therefore became a hugely controversial book because it implied that people were not all born the same, but were biologically suited for particular roles. If differences in people's social position could be attributed to biological programming, the ideals of democracy and racial and sexual equality seemed to be threatened.

Insight

Apart from the controversy, *Sociobiology* was interesting because it systematically examined social behaviour as a biological phenomenon. Wilson explored the chemical 'language' by which different ants within a colony knew what they should do. Their roles were not learned, they were quite instinctive, and could be triggered by giving them an appropriate chemical stimulus. Could human society be similarly explained? Asking such a question has profound political and moral implications.

Within sociobiology, there is a recognition that genes are promoted through the success in breeding of the individuals that carry them. The most obvious example of us acting in line with our genetic programming is in the sphere of sexuality. Males are likely to feel the urge to seek out as many sexual partners as possible, since their supply of sperm is constantly renewed, and the most sensible way of making sure of a good number of progeny is to share it as widely as possible. Although prudence may suggest that it is not ideal to conceive a child with every sexually attractive female within grabbing distance, the fundamental urge to do so is still present.

By contrast, females can only produce a limited number of children. Hence it might be important for them to secure the help of a male that will not only impregnate her, but also provide some measure of protection for her offspring. Traditionally, therefore, males have gone for quantity and females for quality!

One could also examine the sexual taboo against incest on the same basis. If a sexual union is unlikely to produce a child that is genetically viable, then it is biologically discouraged, and this finds its expression in social and religious rules.

Morality steps in and tries to regulate the sexual jungle – but is that not just another attempt to maximize the number of successfully reared offspring? One might argue that, in terms of the survival of society as a whole, some restraint may actually promote the welfare of the next generation.

Insight

There is also a danger that issues within sociobiology will fall into the same trap as the 'tautology' criticism of 'survival of the fittest' (see above). The sociobiological approach is not saying that certain genes 'ought' to survive, nor that 'survival' is what genes are for, simply that – as a matter of observation – those genetic traits that promote the survival and breeding of individuals in a competitive environment will themselves survive.

It may be argued that a distinctive feature of human society is the capacity of individuals to show altruism and co-operation rather than rivalry. The response to this is straightforward. If helping others, or even giving up one's own life for the sake of others, is what one feels impelled to do, that too may be seen as promoting the overall benefit of the genetic pool, since a society where everyone was 'selfish' in the narrow sense might well be self-destructive.

Example

Take the case of a mother who loses her life in the attempt to save her child from drowning. At one level, that appears to be an absolutely unselfish act. But it could be argued that we are genetically programmed to sacrifice ourselves for the sake of the continuation of our genes. Such an act is no more than a cell is required to do when it grows in the wrong place, or is no longer needed. Individuals are expendable, it is the ongoing genetic stream that counts.

Perhaps that example is too obvious, because the child to be saved (whether the rescue is successful or not) is a direct genetic descendant. What is the person drowning is a stranger? Here, one might argue that – within the human species – it is sometimes necessary for individuals to sacrifice themselves for the sake of the whole. Even if it is not their genes directly, it is at least the genetic make-up of the species that is being protected.

Genes are there to preserve and protect their own kind. Our emotional and moral promptings may be no more than the way in which we experience and respond to those genetic promptings. We may feel that we are being rational, independent of our basic biological urges, but sociobiology would challenge that.

Why anyone would throw away their life in order to attempt to save that of an animal is, of course, another matter.

The application of evolutionary principles to human society did not arrive with *Sociobiology*, having been explored by Herbert Spencer in the nineteenth century. What is new, however, is the assumption that all activity is related to genetic traits. It is all too easy to fall into a simple equation of human behaviour and the motivation of the individual with the traits programmed by genes.

A valuable note of caution here is found in Richard Dawkins' *The Selfish Gene* (1976). According to Dawkins, our genes are inherently 'selfish', in that their task is to promote survival and successful reproduction. Our bodies are, from a genetic point of view, vehicles which aid the survival and development of our genetic material. But he cautions that it is not right to see 'selfish' as a moral term here. Dawkins is not saying that genetic theory somehow causes and therefore justifies 'selfish' behaviour on the part of human individuals. The selfishness of competition happens at the genetic level, not the human one.

THE GENETIC ENVIRONMENT

Many ethical issues arise because particular applications of science open up new possibilities that require a balancing of values and motives. Take the debate over genetically modified food crops. Some artificially engineered genetic modifications result in species with beneficial characteristics that could not have developed naturally because they involve the use of genes from very different species that would never normally have been able to cross-fertilize. That is all very well, but there is a risk that such novel species could cross-fertilize with others in the environment, with quite unknowable consequences. The issue here is that of interconnectedness. Genetics has shown the fundamental connectedness of all living things, but genetic research generally seeks to produce a life form that is immune from unwanted influence from its environment. To create a plant that is unaffected by disease may do wonders for one's production of crops, but it is to create a species that is out of step with the surrounding process of natural selection and genetic change.

One way to avoid the problem of cross-fertilization is to create sterile crops that are incapable of producing seed once they have grown to maturity. This preserves the environment, but produces another interesting moral dilemma, for such crops depend upon a supply of new seed each year, since seeds cannot be taken from them and used for planting the next crop. This in effect makes the crop 'copyright', sold to produce a single yield only. This would have great benefit for those who supply seed, but would harm those (particularly in poorer countries) who depend on the ongoing harvesting of seed rather than paying for new seed for each planting.

ESSENCES?

Before Darwin, it was generally assumed that each species had a distinctive 'essence' – the characteristics that made it what it was. Indeed, then (and now) the classification of species has been a major part of the biological sciences. It is also clear that there are features of, say, a horse, that distinguish it from other animals, and it therefore makes sense to say that a horse is a 'natural kind' or that it has a distinctive 'essence.' If that is not possible, what is the point in the classification and naming of species?

The dilemma is that, from the perspective of evolution, we know that species are not fixed, so there can be no definite 'horse', but only a set of individuals of that species that will – over successive generations and a very long period of time – change. We can perhaps think in terms of DNA groups, or some other technical way of defining features of a species. We need some form of dynamic description, allowing each species to be sufficiently described to enable it to be identified, while taking into account the process of modification over time.

Insight

After all, the title of Darwin's book – *On the Origin of Species* – is exactly that, it is about 'species', which implies some form of classification by way of essence, albeit an essence that is developing rather than fixed or imposed.

KEEP IN MIND...

1 The debate about the appearance of 'design' in nature predated Darwin.

2 Natural selection is an observation of what actually happens in nature.

3 The theory of natural selection explained a mechanism for change, parallel to the artificial process of the selective breeding of domestic animals.

4 'Survival of the fittest' really means 'survival of those best fitted to their environment'.

5 Genetic mutations provide the source of the small differences that are required for natural selection to operate.

6 Genetics shows the interconnectedness of species by identifying genes with similar functions in different species.

7 Genetics shows that each individual is unique but that all are connected.

8 Key question: 'To what extent are we the product of our genetic inheritance, as opposed to our environment and upbringing?'

9 Sociobiology applied the principles of molecular biology to animal and human society.

10 Classification implies essence, but evolution shows that there is no fixed essence.

9

··

Cosmology

In this chapter you will:

- *consider the ability of science to describe the origins and nature of the universe*
- *examine the attempt to achieve a single explanatory theory*
- *explore the human perspective on the universe.*

Understanding the nature of the universe as a whole has always been a central quest for both science and philosophy. As we saw in Chapter 2, the rise of modern science, with its emphasis on observation and experiment and its use of mathematics, promoted a new approach to astronomy. As we look at the work of Copernicus, Galileo, Kepler, Newton and others, we see both the quest to understand the structure of the universe, but more specifically to discover the laws of physics which would account for the movement of bodies, both heavenly and terrestrial.

In cosmology, there are two interrelated sets of questions:

1 The first question

> What is the structure of the universe? How did it originate? What is its future? How did it develop its present form? By what physical laws can we understand its workings?

▶ Questions of this sort have led us from the Ptolemaic Earth-centred universe, through Copernicus to Newton, and on to the modern image of the world expanding outwards for the last 15 billion years from a **spacetime singularity**. They are concerned with order, structure and the process of development.

2 The second question

What is the simplest, most basic and most general thing we can say about reality? What lies beneath the multiplicity of what we see?

▶ Questions of this kind started when Thales speculated that the world was essentially composed of water, and the atomists tried to find the basic building blocks of physical reality. They were implied within the world of Newtonian physics by the quest for ever more general and comprehensive laws of nature. More recently they have been expressed in the quest for a single 'theory of everything' that might show how the fundamental forces of the universe (electromagnetic, gravitational, strong and weak nuclear) relate to one another.

In modern cosmology, these two sets of questions come together. The structure, origins and development of the universe are closely linked to questions about the matter that is within it, to the point of needing to infer the existence of unobservable, dark matter in order to account for the universe's present form. If we get answers to one set, they will inform our understand of the other.

Dimensions and structures

Our galaxy is spiral in shape and rotates. It is thought to contain over 200 billion stars, and to be about 100,000 light years in diameter. The Sun is a smallish star, about 32,000 light years from the centre of the galaxy. It is estimated that there are at least 100 billion galaxies. These are not spread evenly through space but are clustered.

Insight

Since the earlier edition of this book, written nine years ago, the estimated number of stars in our Milky Way galaxy has doubled, and the total number of galaxies in the universe has increased by a multiple of 10! By the time you read this, even these figures will be out of date.

Since it takes light a considerable time to travel from one part of the universe to another, we are looking back in time as we look out into space. If I observe a galaxy 5 million light years away, what I am actually observing is the state of that galaxy as it was 5 million years ago, when the light from it started its journey. The light from galaxies more than 5 billion light years away (which is still less than half way across the known universe) started to travel towards me at a time before the Sun or the planets of our solar system were formed. If an observer on that galaxy were looking in this direction today, he or she would see only dust clouds.

The dimensions of the universe are equally astonishing in terms of emptiness. We tend to think that the Earth is solid and compact, but in reality it is, of course, mostly empty space, through which particles can pass unhindered.

The mass of neutrinos?

In June 1998 a team of 100 scientists from 23 different institutions in America and Japan announced the results of an experiment to detect and measure the mass of neutrinos, elementary particles so small that they are capable of passing through the Earth. Their work involved a tank of very pure water, one mile below the Earth's surface, in which, once every 90 minutes, a neutrino made its presence known by colliding with an oxygen atom and giving off a flash of blue light.

The project involved looking at the secondary particles that strike the Earth, following the bombardment of the upper atmosphere by fast-moving particles from space. It was important for an understanding of the structure of the universe because previously it had been assumed that neutrinos had no mass. This had implications for a big issue for cosmology at that time – namely, whether the mass of the universe was such that its gravity would prevent it from expanding indefinitely, and the suggestion was that neutrinos might account for much of the missing (or 'dark') matter.

It is curious to think that the universe may take the form it has, and have its future determined by particles so small that they can pass through the Earth undetected. It highlights the very narrow range of objects that humans generally regard as having significance. Visible matter is seen as significant simply because it is visible; other levels of material reality remain hidden, but are equally important.

Wind forward, and we know now that the universe is expanding at an ever-increasing rate. But experiments in 2011 appeared to show that neutrinos could travel faster than the speed of light, which appeared to contradict Einstein, unless of course the neutrino had no mass (since Einstein's prediction that nothing could travel faster than light was based on the assumption that it would take an infinite amount of energy to get anything with mass beyond that speed). So does that reopen the question of whether neutrinos have zero mass? In any case, it is likely that the mass of a neutrino is perhaps less than a billionth of that of a hydrogen atom.

Insight

I do not know how this issue will be resolved, nor – like most people – do I have the expertise to understand or judge the outcome. But it serves to illustrate the interlocking nature of the two sets of questions with which this chapter opened. The whole structure of the universe may hang on the weight of a particle.

Einstein suggested that the universe could be seen as a hypersphere, and that one could move through it indefinitely ever without reaching an edge, even if it were in fact finite, eventually coming back to the point from which one started. This sounds bizarre, until one considers that General Relativity showed that space and time are not fixed; space is bent in strong gravitational fields and gravity is one of the fundamental forces that holds the universe together. Thus, all space is going to be bent to some very minute degree; and if space is bent, then it will eventually fold back on itself.

One way you might envisage this infinite distance through a finite universe is in terms of moving round the inside surface of a sphere. You could turn any way you liked and travel an infinite distance on that surface, but the surface itself remains finite. However far you travel, you will never be further away from any other point on that surface than the diameter of the sphere. (Ever watched a hamster on a wheel or in a ball?)

However, Einstein assumed that the universe remained the same size. Why then did gravity not cause it to collapse back in upon itself? His answer was to propose a 'cosmological constant', representing a force that held things apart, frustrating gravity. Later, when it was shown that the universe was expanding and that the constant was therefore unnecessary, he admitted that it had been a blunder.

THE 'BIG BANG' THEORY

In terms of terrestrial experience, you can observe only the present,
never the past. The past is, by definition, what no longer exists. In
terms of the universe (because of the limitations imposed by the speed
of light), the opposite is true. You cannot observe the present, only
the past. The further you look, the further back you see.

One way of knowing about the past is to observe trends in the
present and extrapolate back from them. We can observe that
galaxies are moving apart, both from one another and from us. But,
since the spectrum of light changes as objects move away at very
high speeds, we also know (as a result of the work of E. P. Hubble
in 1929, who observed a 'red shift' in the light coming from distant
galaxies) that the further away from us a galaxy is, the faster it is
moving away. Clearly, therefore, if the galaxies are moving apart in
this way, the universe is expanding.

From the speed of expansion, it is possible to calculate the age of the
universe. About 13.7 billion years ago the present universe started
expanding outwards in an 'explosion' known as the 'hot big bang'
from a point where all space, time and matter were compressed into
an infinitely small point, called a **spacetime singularity**. A crucial
thing to appreciate about this theory is that a singularity is not a
point within space and time – it is the point from which space and
time have come.

Expanding space

It is difficult to think of an explosion without thinking of matter
flying outwards through space. But, in this early stage, there was
no space 'out there' through which matter could explode. Space
is created as this expansion takes place. It is also confusing to say
that the universe started from a point (a singularity) that was

very small. Imagining a universe the size of a pea is confusing because we automatically imagine a world outside that pea, and yet (according to this theory) there is no 'outside'.

The expansion of the universe is not 'through' space, but 'of' space. This is difficult to conceptualize, since – on the small scale seen on Earth – we see space as static; a metre rule remains one metre in length. Yet, if space is expanding, then so is everything. The same process which propels distant galaxies away from us is also gradually pulling atoms apart, and extending metre rules! In other words, there is no point in space from which this expansion is moving out; it is moving out from all points simultaneously. The 'Big Bang' did not happen somewhere else; you are part of its expansion.

The process by which the universe takes its present structure is one in which energy becomes matter, spreading out uniformly in the form of hot gas. This then cools and condenses, gradually forming the galaxies.

However convincing this argument, it is still a projection of present trends back into the past. Yet the past is visible, if we look far enough. In theory, therefore, if the universe did expand out from this 'Big Bang', then we should be able to look back through time and find some trace of it. And the key to that, of course, is that – if space has 'grown' with the expansion – then the traces of the 'Big Bang' are not going to be located in any one place, but will be spread uniformly across the universe.

Such direct evidence was provided in the 1960s, when background microwave radiation at a temperature just above absolute zero was found throughout the universe.

Every theory is confirmed or refuted on the basis of the things that it predicts – in other words, we have to ask what follows from it, and then check out whether that is the case. In the case of the Big Bang theory, several quite obvious consequences have been checked:

▶ First of all, we can look for some evidence of that early state. That was found in the background radiation.
▶ Secondly, if the universe started as a Big Bang – in other words, in a state where everything was flying apart – then (unless gravity

had been strong enough to halt that expansion at an earlier stage) it should still be expanding today. This has been confirmed by the observation of the 'red shift' in distant galaxies.

▶ Thirdly, if the universe expanded as predicted by this theory, there should be substantial quantities of the light elements, particularly hydrogen, in the present universe. The quantities of these elements that are observed today are in line with the quantities predicted by the theory – the universe is 75 per cent hydrogen.

There are many outstanding problems here. One is to find out where the vast amount of matter and radiation in the present universe 'came from'. How is it that a sudden expansion of this sort comes about? Looking at the possible trigger for the Big Bang, the theory developed by Neil Turok (1958–) and Steven Hawking (1942–) sees its origins in what they term an 'instanton' – a point that includes space, time, matter and gravity. It lasts for no more than an instant, but has the ability to trigger the production of an infinite universe.

What is more, long before the galaxies were formed, we know that there was an unevenness in the universe, and unevenness that may have played a crucial role in determining how the hot gas condensed to form the sort of structures we see today.

Work over the last two decades on the 'inflation' model describes a state in which there is a very rapid expansion along with the spontaneous creation of matter and energy. This is thought to have happened during a very brief early phase of the Big Bang. Slight quantum variations in this inflationary stage might explain the later ripples of unevenness in the expanding universe, which in turn caused the formation of galaxies as the universe cooled.

Insight

Whether it is the ripples in the early universe, or the quest for the elusive 'Higgs boson' which is thought to enable particles to gain mass, the problem for the non-specialist is how to conceptualize what is being described, and to grasp how it is that the structure of the universe is intimately bound up with the nature and behaviour of subatomic particles.

There is always going to be a limit to the visible universe. Galaxies that are moving away from us at a speed at or greater than the speed of light will be invisible to us, since – given that the universe started at some point in our time frame – there has not been enough time for their light to reach us. That curious fact does not deny Einstein's claim that matter cannot travel faster than light, since that is based on the idea of space being static, whereas if space is expanding, there is no such limit. Whatever the size of the universe, there will always be an absolute limit to how much of it we can experience.

WINDING DOWN OR BOUNCING?

The Second Law of Thermodynamics describes the loss in terms of heat energy in every change of state. In other words: in any closed system, things gradually wind down. Applying this to the universe suggests that it should gradually dissipate its energy and end up in a state of total entropy (uniform disorder) having lost all its heat.

The outward momentum of the galaxies is countered, of course, by the gravitational pull exerted by their mass. A closed universe is one in which the expansion is slowing, due to the force of gravity, and will eventually re-collapse into itself. An open universe is one in which there is not enough matter to enable gravity to halt the expansion.

This led to a discussion about whether the mass of the universe was such that gravity would eventually halt its expansion and lead to a 'big crunch', which might in turn lead to another Big Bang – in other words that the universe was bouncing! It also led to the prediction, since the visible matter in the universe was insufficient to account for the amount of gravity, that there must be 'dark matter', invisible to us but exerting gravitational pull.

From observations of supernovae, made in 1998 (for which Perlmutter, Schmidt and Riess were finally given the Nobel Prize in Physics in 2011), it was finally confirmed that the universe has been expanding at an increasing rate, so the 'big crunch' scenario appears to be eliminated. But, just as the amount of gravitational attraction suggested 'dark matter', so an accelerating universe requires 'dark energy' exerting a negative pressure, to account for this increasing rate of expansion. We therefore appear to be living in an open universe.

Notice that all theories are liable to change over time, in cosmology as elsewhere. Dark energy and dark matter are required in order to fit our present theories to existing physics. But keep in mind an interesting historical precedent: Newton introduced the idea of the 'ether' because he could not understand how gravity could operate over a distance without some connecting medium. It took a long time for the scientific community to realize that the 'ether' was not needed as an explanation.

Towards a theory of everything

Generally speaking, we have seen that science makes progress by formulating theories that account for existing evidence and which can be used to make predictions. As theories come into contact with one another, they either clash (in which case the more productive and successful at prediction tends to displace its rivals) or agree (in which case each is reinforced by the other).

We have seen also that, as one general paradigm gets replaced by another, it is often because the earlier one was limited in its application. So, for example, Newtonian physics was shown to be fine, but only for the conditions found on Earth, whereas Einstein's theories could be used to explain more extreme conditions in phenomena such as black holes.

Logically, therefore, the process of refining and developing scientific theories should be capable of moving in the direction of a single theory able to account for all phenomena everywhere: TOE – a 'theory of everything'.

There are four basic forces: the strong and weak nuclear forces (which operate over short distances and hold atoms together), electromagnetism and gravity (which is a far weaker force, but operates over vast distances giving shape to the whole universe). A TOE would seek to reconcile all four within a single theoretical framework.

To test the relative strengths of the nuclear and gravitational forces, it is only necessary to fall over. Gravity will have the more obvious effect over a distance of about a metre or so, as you topple and fall. On the other hand, at the moment of impact, the nuclear forces holding together the atoms that constitute the ground and your body are clearly superior over that very short distance, and bring you to a sudden halt!

The further you go back in time towards the singularity, the simpler the universe becomes. Once matter is formed from radiation and starts to move apart, the fundamental forces that are later to shape it differentiate from one another. Yet, at the point at which the whole universe is compressed into a single point, those forces cannot be separated. Just as, in the area of biological evolution, one may look back to a common ancestor for two related species, so in the area of fundamental physics, one may look for a common ancestor for the fundamental forces. It is a quest for simplicity rather than complexity, reflecting the way in which the universe has moved from a very simple early state to a more complex one now.

The problem with describing a very small, compressed universe is that it is unimaginable. The most significant feature of a small-scale object is that its structure is simple, undifferentiated. Therefore, instead of imagining size, we need to imagine structure. The simple, extremely hot, early universe gradually cools, and matter forms, is differentiated and expands outwards, clumping together to form galaxies under the influence of gravity.

Insight

A theory of everything should, if it reflects reality, be capable of predicting everything. In other words, once we understand how the four fundamental forces can be united – as they must have been in the earliest universe – then it should be possible to capture, in mathematical or logical terms, how it is that a universe develops. But just as quantum mechanics has shown that it is impossible to measure both the velocity and position of a particle simultaneously, so we may well find that there are insuperable problems to achieving a single TOE.

In one important respect, cosmology is different from other branches of science. There is only one universe, and this obvious fact is reflected in the way cosmology goes about its task. If you want to understand, say, the operation of the brain, you can set up various experiments using human or animal subjects, looking at the way in which parts of the brain relate to sense experience, reasoning or emotion, or activity. You can compare the way in which the same function is handled in different species. You can isolate the variables you want to examine, set up appropriate experiments and, if necessary, go looking for more evidence.

That, of course, is impossible for cosmology: we cannot compare different universes. Theories about the origin and nature of the universe, based on evidence (e.g. galaxies moving apart) are put forward and examined, generally using complex mathematics. We can ask: 'What if there were more or less matter (and therefore more or less gravity) – what sort of universe would result if that were the case?'

As a result of doing this, theories are put forward and entities sometimes posited in order to account for the way the universe is observed to be. If there is insufficient mass to account for the amount of gravity, we hypothesize that there is 'dark matter', exercising gravitational pull but undetectable by us. If the universe is expanding more quickly than is expected, we suggest that there may be 'dark energy' forcing matter apart.

Such hypotheses – like Einstein's mistaken use of a 'cosmological constant' – are needed in order to make our observations fit our assumptions. In Einstein's case, he assumed the universe remained the same size, and needed his 'constant' in order to achieve that.

Our present ideas about mass and gravity require 'dark matter' or 'dark energy' in order to fit them to the observed universe – but we may eventually conclude that it is our assumptions about mass and gravity that are wrong. That is all part of the challenge of science.

The human perspective

The study of cosmology generally raises issues that go beyond the search for a mathematical theory to explain the origin and development of the universe. It does not remain simply a quest for information, but inevitably raises questions about the meaning and value of human life. The natural thing to want to ask about any general theory of the universe is where and how human life fits into it, and whether it can have any lasting significance. Such questions are likely to be philosophical and religious rather than scientific. But they can sometimes influence our interpretation of the evidence.

One of the issues that we have faced time and again within the philosophy of science is how to take into account the fact that what we observe is influenced by our own faculties of observation. Our understanding of the world is influenced by the way we examine it and the questions we consider it appropriate to ask of it. But there is a more general point about our observation of the world – the fact that we are here to observe it! The question is this:

What is the significance of the fact that we know the universe is such that it can – and has – produced human life? This is explored in what is known as the 'Anthropic Principle'.

THE ANTHROPIC PRINCIPLE

Imagine a universe in which one or another of the fundamental constants of physics is altered by a few percent one way or the other. Man could never come into being in such a universe. That is the central point of the anthropic principle. According to this principle, a life-giving factor lies at the centre of the whole machinery and design of the world.

J.D. Barrow and J.A. Tippler, *The Anthropic Cosmological Principle* (1986)

There are two versions of the Anthropic Principle: a weak one and a strong one.

1 **The weak version** If any of the major constants of the universe were different, we would not be here, life would not have evolved. Our life is dependent upon the world being exactly as it is.
2 **The strong version** The universe contains within itself the potential for life, and that it was therefore impossible for human life not to have been created in this world.

Let us examine the logic of this argument. It would seem to be this:

▶ If the universe were different from the way it is, we would not be here.
▶ Everything that happens is determined causally.

▶ Once the initial parameters of the universe were established, it became quite inevitable that life would evolve, and therefore that we would end up contemplating the universe.

This raises a fundamental question. We know that the initial conditions must have been what they were in order for this kind of universe to develop, and therefore for us to be here. But did the universe *have* to be like this?

Now, the strong version of the Anthropic Principle makes it sound as though there was an element of compulsion in the early conditions of the universe; in other words, that they had to be what they were, and that the universe therefore had no option but to develop life on this particular planet.

But this goes against the whole descriptive nature of scientific theories and laws, namely that they are the summary of experience, expressed in the form of a general proposition. They are either accurate or inaccurate in describing the world – they do not command the world. You cannot 'break' a law of nature. Laws of nature do not determine what shall happen; they summarize what has happened and thereby (if they are successful) may be used to predict what will happen.

But in its weak sense, the Anthropic Principle says, in effect, that everything depends on everything else, and that, if anything were different, then everything would have to be different. In a different universe, we would not be here. That is no more than was claimed by Leibniz – who saw the whole world as a mechanism in which a change in one place implied changes elsewhere.

Insight

It is obvious that, if the conditions had not been right for humankind to have appeared, then humankind would not have appeared; but that gets us nowhere in terms of understanding the original conditions of the universe.

We know, of course, that any explanation of the universe which would not have allowed the formation of galaxies, for example, could not have been correct. In that sense, we are reminded by the Anthropic Principle that we (and the conditions that made us possible) are part of the evidence that any theory needs to take into account.

'We are part of the universe.' This banal statement has enormous implications for an overall view both of cosmology and science as a whole. From the time of Descartes through to modern discussions of the Anthropic Principle, there has been a danger of separating humankind off from nature, seeing the world as something 'out there' to be understood, or as a process whose sole purpose is to produce thinking human beings.

Seeing human beings as somehow 'outside' the natural world, and treating the latter as little more than a vehicle and life-support system for their own benefit, leads to a distortion that we see frequently in terms of our terrestrial environment and the impact that humankind has upon it.

One of the features of the cosmos that inclines people towards an acceptance of the Anthropic Principle is the unlikely sequence of events that must have taken place in order for human (or any!) life to evolve on Earth. Once we examine the facts, however, we come across the old problem of chance and necessity:

On the one hand, everything is extremely improbable, in that it depends on a huge number of other factors being the case, but on the other, given those factors, it seems inevitable.

Take the example of planet Earth. Life is sustained on this planet because conditions are just right, and in particular because it is a wet planet, with much of its surface covered with water, which is taken up into the atmosphere and deposited again as rain. Without water, there would be no life as we know it. Looking at neighbouring planets we see that life could not be sustained, and we realize that it is only because our planet is a certain distance from the Sun that water can exist in liquid form, and everything that follows from that simple fact.

However, looked at in a different perspective, water is very common, and is likely to be in liquid form on any planet that is orbiting at roughly this distance from its star. Far from being a unique case, the Earth may be simply one of a huge number of wet planets, spread throughout the galaxies – each with oceans, clouds and rain. If the elements of which we are composed are those found throughout the universe, why should we assume that the Earth is so special?

Getting the atmosphere right

At one time it was generally believed that the early atmosphere on the Earth was composed of huge quantities of methane and ammonia and thus not compatible with the development of oxygen-needing forms of life, but it was argued that, if you take methane, ammonia and water, expose it to ultraviolet radiation, and then pass electric arcs through it, molecules are produced that will eventually start moving in the direction of life. Those required conditions would have been found on Earth, since the Sun would have provided the ultraviolet light, and thunderstorms would have given rise to frequent bolts of lightning.

A more likely scenario (e.g. as set out in the very accessible collection of articles *The Case of the Missing Neutrinos* by John Gribbin) is that the atmosphere was at one time largely composed of carbon dioxide, some of which was absorbed by the oceans. As life started to develop in the waters, oxygen was produced as a by-product, and this then built up, gradually changing the balance of gasses in the atmosphere. The other key change was the development of the ozone layer, which cut ultraviolet radiation from the sun and enabled the further development of life on dry land, which would previously have been destroyed by the radiation.

Whether either of these theories is correct is not the issue. The point is that the appearance of any one form of life depends not just on place but on time; human life fits into a sequence and survives for only as long as conditions permit. That highlights again the unlikelihood of human life, and the paradox of seeing how well the universe and conditions on Earth appear to have been prepared for the appearance of humankind, while recognizing the limited validity of taking such a retrospective view.

Whatever else it reveals, astronomy is a science that is supreme at putting human concerns into perspective. Our Sun is already well through its life cycle, and one day it will expand, destroying the Earth in the process. Even that, however, is merely a temporary and parochial incident compared with the future of the galaxy.

Our own Milky Way and the Andromeda Galaxy are approaching one another at 300,000 miles per hour. They will eventually collide, either head-on or grazing against one another. This will start to happen in approximately 5 billion years' time, and the process will probably take several hundred million years, involving the birth of millions of new stars as huge molecular gas clouds are compressed between the galaxies. Eventually a new composite galaxy will form, and all that we see now of our own galaxy will be totally transformed.

Insight

Such is the nature and dimensions of the universe, of course, that, long after the material that presently forms the Earth, Solar System and galaxy has been recycled, it will be theoretically possible to view it – just as it is today – from some distant point in the universe.

What is clear, however, is where we stand in this process. The first generation stars were formed about 10 billion years ago, fuelled by nuclear fusion, as quantities of hydrogen were turned into helium, and other heavier atoms were formed. After star death in a supernova, the resulting matter was scattered deep into space, gradually gathering into clouds which condensed under gravity to form another generation of stars and planets, ours included. Every atom in our bodies was formed out of hydrogen inside a first-generation star.

From the time when energy turned into matter, and matter clumped together, fuelling heavier atomic structures, a process has been going on of which we are a part. This is not to say (as the strong Anthropic Principle tends to suggest) that the whole process was there for our benefit, designed in order that we could be produced. Rather, it is to acknowledge that we are a very small and very temporary incident in the ongoing cosmic story.

Insight

It is instructive to contemplate the fact that 75 per cent of the material in the universe is hydrogen, 24 per cent helium, and all the heavier elements make up the tiny remaining portion. We are part of a rare exception!

KEEP IN MIND...

1 The structure and origin of the universe are related to the most general and fundamental features of energy and matter.

2 Cosmology is directly related to work on the nature of subatomic particles.

3 Einstein originally assumed that the universe remains the same size and therefore proposed a 'cosmological constant' to account for this, later regretting his mistake.

4 The universe originates in a spacetime singularity, where space, time and mass are compressed into an incredibly small space.

5 Space expands *with* the universe; the universe does not expand *through* space.

6 There are four basic forces: strong and weak nuclear, electromagnetic and gravity.

7 A 'theory of everything' seeks a single explanation unifying these four forces.

8 Unlike most other areas of science, cosmology is limited by the fact that there is only one universe, so it cannot compare or test different versions.

9 The Anthropic Principle claims that the universe needs to be such that we are able to exist to observe it.

10 Cosmology highlights the number of factors that are needed in order to make human life possible.

10

Science and humankind

In this chapter you will:
- *consider to what extent humans can be regarded as machines*
- *explore the implications of neuroscience*
- *examine the social functioning of science.*

Most of what we have been considering so far relates to how human beings understand the world around them. Science is concerned both with what is 'out there' to be known, and with the process (through the action of the senses and the mind) by which we come to know it. So any understanding of science requires an epistemology (a theory of knowledge) and also an appreciation of how the mind works.

From the time of Galen (129–210 CE), who combined the work of a doctor with that of a philosopher, the human body and mind have been the object of scientific study. In this chapter, therefore, we shall outline some of the issues that arise when human beings use science to understand themselves.

Prior to the nineteenth century the discussion was of the nature and operation of the physical body, and of the relationship between mind and body – as well as the more general philosophical issues of human behaviour, in ethics and politics. But since the nineteenth century the range of scientific approaches has expanded to include psychology, sociology and – with the impact of the theory of natural selection – the nature of humankind and its relationship to other species.

Alongside this (although beyond the scope of the present book) we need to keep in mind that science and technology have taken an increasingly active role in shaping human wellbeing and experience. Information technology has transformed social communications, medicine has prolonged life expectancy and technology has (for good or ill) transformed the environment. We cannot understand humankind today except in the context of technological progress made possible by science. It represents the most obvious way in which *Homo sapiens* is different from other species.

The human machine

Although writing in the second century CE, **Galen** was quite capable of examining parts of the body and considering their function in terms of what they contributed to the whole. Sometimes, of course, he was wrong. For example, he assumed that the purpose of the heart was to create heat, and that the lungs drew in cool air in order to stop the body overheating. However, he established an important principle for an examination of anything as complex as the human body, namely that it had to be considered as a whole, and that the parts were there to serve some overall purpose.

His views on the heart were overturned when, following years of experimentation, **William Harvey** (1578–1657) published *Exercitatio anatomica de motu cordis et sanguinis in animalibus* (An Anatomical Exercise on the Motion of the Heart and Blood in Living Beings) in 1628, establishing that the heart was a pump, and that blood was oxygenated in the lungs before being pumped through the body. This enabled progress to be made in analysing the workings of the body, for it became clear that each organ was nourished by oxygenated blood, and that the whole body was thus an interconnected system.

In common with other features of the science of the seventeenth and eighteenth centuries, Harvey's work on the human body showed that it, like the universe as a whole, could be seen as a machine. At the same time, Descartes was working on his philosophy that was to separate the unextended mind from the extended physical body, with the latter able to be examined like any other machine, controlled by mechanical laws.

We have already considered the contrast between a reductionist and a holistic approach to the understanding of complex entities (see Chapter 5). The crucial issue for an understanding of the human being is whether it can be adequately explained in terms of the mechanisms that are revealed by a reductionist analysis (in other words, whether the brain, nervous system and so on explain human behaviour). If it can, then human beings can be regarded, from a scientific point of view, as machines – albeit the most sophisticated machines imaginable.

Insight

This is not to denigrate the importance of reductionist analysis. If I am ill, I may well need a blood test or biopsy, both of which consider the implications for my health by analysing a small portion of my body. The question is whether the findings of such analysis could, in theory, give an adequate account of my life as a whole.

A key question is whether science can examine humankind just as it examines other phenomena:

To what extent is it possible for the human mind to explain its own operations?

Later in this chapter we shall examine this in terms of the status of social and psychological theories, and particularly in work done in neuroscience. In examining these, we need to keep in mind a fundamental question:

Is the 'machine' model adequate for understanding humankind?

This is not an easy question to answer because in some respects, particularly in medicine, the machine model has served us well. However, just as the experience of freedom is difficult to reconcile with the view that everything is determined by antecedent causes, so the idea that psychology or neuroscience can give an exhaustive account of my mind may feel improbable, even if reason cannot refute the idea.

Human origins and evolution

In terms of human self-understanding, Darwin's *On the Origin of Species* must rank as an absolutely pivotal work. Evolution gives us a new perspective on humankind, linking it to all other living beings. Since the nineteenth century there have also been considerable advances in knowledge, through the discovery and analysis of early remains, of the development of our own species over what is (in evolutionary terms) a very short period of time.

The Earth was formed about 5 billion years ago; dry land and then primitive vegetation developed about 410 million years ago. So, for most of its life, the planet has been watery and lifeless.

The progressive evolution of living things has been punctuated by major destructions at 250 million years (with approximately 90 per cent of species destroyed) and 65 million years ago (most famous for bringing to an end the era of the dinosaurs, but destroying approximately 50 per cent of other species as well). Mammals appeared only 50 million years ago, and the first apes 35 million years ago.

By 10 million years ago there must have been a wide variety of ape species, including those who were to develop towards present-day gorillas, chimpanzees and humans. Most of these species died out, and little is known of them, but the molecular clocks in DNA give a clear indication of when species separate off from one another. We therefore know, by comparing the DNA of chimpanzees and humans, that the two species diverged about 7 million years ago, and there emerged in Africa a family of species called the hominids.

In 2002 the skull of an early hominid, nicknamed 'Toumai' – meaning 'hope of life' – was discovered in northern Chad. It is over 6 million years old and comes from a time when hominids must have been very rare. They emerged out of the general range of ape species, and over time all but one of them (which was to become *Homo sapiens*) died out.

Insight

In trying to imagine the world of 7 million years ago, it is important to remember just how insignificant the hominids must have been among the whole range of animal life. It has only been during the last 10,000 years that humans have multiplied in numbers sufficient to dominate the planet and finally to claim (whether rightly or not) priority over other species.

Australopithecus (found in Ethiopia) dates from around 4 million years ago (with a brain capacity increase from 450 cc (cubic centimetres), found in the early apes, to around 750 cc).

One million years ago *Homo erectus* is found in Africa, Asia and Europe (the brain capacity increasing from 800 to around 1,200 cc), already shaping stones and using fire, and recent findings in the Republic of Georgia are dated back between 1.7 and 2 million years.

Neanderthal Man probably developed up to 300,000 years ago (according to dating of Neanderthal remains in Spain), and by 75,000 years ago had burial places and funeral rites. The branch of the human family is generally thought to have gone extinct about 32,000 years ago.

Homo sapiens appears in eastern and southern Africa, about 100,000 years ago (or perhaps earlier), and by now its brain capacity has reached around 1400 cc, much as it is today. It is assumed that it moved out from there to colonize the other continents, but there is disagreement about how that happened.

The development of humankind becomes more complex, since the discovery of a smaller 'hobbit' form, named *Homo floresiensis*, on the island of Flora in Indonesia, which is thought to date from about 700,000 years ago. The suggestion therefore is that *Homo erectus* had crossed to this part of the world long before *Homo sapiens* had left Africa, perhaps as early as 1.6 million years ago.

Rock art, thought to be 75,000 years old, found in Australia also suggests that it was *Homo erectus*, rather than *Homo sapiens*, who made its way there. If so, crossing at least 40 miles of ocean would have required speech and social organization of some sort, features that were thought not to have developed until *Homo sapiens*.

By around 30,000 BCE we find cave paintings in south-west France (although, of course, if the Australian claims are correct, European culture was late developing), and by 10,000 BCE you have settled communities within the 'fertile crescent' of the Middle East. And the rest is history!

Insight

This is no more than a brief outline in order to get the timescale into perspective. New discoveries are constantly causing us to rethink the way in which early man developed and spread. The details may be fascinating, but for the philosophy of science we simply need to look at the process of interpreting evidence.

INTERPRETING EVIDENCE

Notice the scientific methodology used in examining human origins. First of all, we have archaeology, and the problem here – as with the more general issue of the fossil record of evolutionary change – is that there is very little physical evidence upon which to build theories.

A small piece of conflicting evidence may be enough to threaten an entire theory of human development and migration.

Here we need to reflect on the issues examined in Chapter 4. A straightforward process of falsification might suggest that a single stone tool is enough to require that a theory be abandoned (on a simplified version of Popper's 'falsification' approach). On the other hand, it might be more prudent to put the contrary evidence 'on hold' to see if it is confirmed by later findings.

Secondly, we need to keep in mind the potential problems involved with the human interpretation of humanity. We may bring our own understanding of what a tool is, or what it takes to cross a stretch of water, as a model for understanding the past – and thus read into the evidence more than we should.

Once finds of human origins are made and dated, there is analysis of, for example, the size and shape of a skull in order to yield brain capacity. Forensic examination of remains can yield information related to diet, body shape, posture and so on. A comparison of these then goes to build up developmental patterns – for example, the increase in brain capacity over time.

Such examination can build up a picture of the life of early humans. One danger in this process is the assumption of a neat and orderly development from one form to another. Thus, for example, it may be assumed that Neanderthals were very unsophisticated compared with *Homo sapiens*, who would have lived alongside them and eventually came to replace them in Europe. But is that justified on the available evidence, or an assumption based on the final domination of the latter? There have been suggestions that Neanderthal life may have been rather more sophisticated than we originally thought.

EVOLUTION

From the start, it was clear that the theory of natural selection had important implications for human self-understanding, and most of the early controversies were related to this. In *The Descent of Man* (1871) and *The Expression of the Emotions* (1872) Darwin suggested that mental ability and social behaviour could have the same form of evolutionary development over time as the physical body.

This is generally termed 'Social Darwinism', an approach originally developed by **Herbert Spencer** (1820–1903), whose own theory of evolution was based on Lamarck's idea that one could inherit characteristics that had been developed by one's parents, which was to be superseded by Darwin's theory of natural selection.

According to Spencer, it was only natural that human society should follow the struggle for survival that went on throughout nature. In America he became hugely popular for his advocacy of free competition in a capitalist system; success and failure were part of the struggle to evolve. In social and economic terms, one should not feel guilty about succeeding at the expense of others, for this was the pattern throughout nature. To act in any other way was to stand in the way of progress. In Britain, he opposed the Poor Laws and state education, on the grounds that it gave benefit to those who were least able to take care of themselves, and thus upset the natural competitive balance in society.

The general criticism of this approach is that it attempts to make what *does* happen the logical basis for saying what *ought to* happen. This is an issue for ethics, rather than the philosophy of science, but briefly (following arguments put forward by David Hume and, at the beginning of the twentieth century, by G. E. Moore) it is held that you cannot logically derive an 'ought' from an 'is', and that a description, in itself, is no basis for a moral prescription – which is known in ethics as the 'naturalistic fallacy'.

The status of social and psychological theories

In Chapter 7 of *On the Origin of Species* Darwin turns from his consideration of the physical properties of creatures to examine their instinctive patterns of behaviour. He argues that social activity, as he observed it in nature (e.g. in the ways ants work together, differentiating their functions within their community), could be accounted for using his general theory of natural selection. He argued that there was a 'slave-making instinct' that led individuals within a colony to act in a way that enabled personal goals to be set aside in favour of the goals of the colony as a whole.

Now it is clear (although Darwin did not take this to its logical conclusion) that humankind can be examined in exactly the same way. We see that society is organized in such a way that individuals sometimes need to sacrifice themselves for the good or the whole or, at the very least, they have to accept a particular role placed upon them by society.

Morality tends to follow from social differentiation, as we see when people feel instinctively guilty if they fail to do what is expected of them. Hence, much of what has been reserved for the personal, moral and religious aspects of life (and therefore previously largely untouched by the arguments and experimental methods of the physical sciences) can be seen in terms of the theory of natural selection.

This line of thinking led to the development of sociobiology (see Chapter 8), and its discussion about the degree to which our behaviour is genetically prompted in ways which reflect the basic genetic impulse to facilitate survival and breeding.

We find that, as science moves from a consideration of physical entities to look at the nature of humankind and society, we are likely to encounter a whole raft of problems, since the traditional division, with the personal, moral or religious aspects of life on one side and the physical and scientific on the other, can no longer apply in a way that is clear-cut.

We saw that in the nineteenth century, as a result of the availability and analysis of statistics about human behaviour, sociologists and others started to examine human behaviour in scientific terms, looking for law-like patterns and regularities within, for example, the incidence of suicide in the population. As a result, we see the development of studies of humankind that claim to be scientific, in that they follow the basic method of inductive inference.

Hence you have the development of political science, sociology and psychology – all of which examine aspects of human life that had not previously been within the remit of science. Now, clearly, philosophy is concerned with all of these, since they raise a whole raft of issues. For our purposes here, we will keep strictly within issues already raised within the philosophy of science and examine the extent to which this new science of humankind can be said to be 'scientific', in the sense of keeping within the parameters of accepted scientific method. To do this, we will look at two of the best known exponents of such new thinking: Marx and Freud.

KARL MARX

Karl Marx conducted extensive historical research into the causes of social and political change. He observed that human survival depended on the supply of food and other goods, and that the function of society was to organize the production and distribution of these. He therefore pictured society in terms of networks of relationships, based on the fulfilling of human need. He saw all other aspects of society as shaped by this economic infrastructure.

A scientist starts with evidence and seeks to find a theory that will explain it and also serve as a basis for predictions. The acceptance of the theory will depend upon whether those predictions are subsequently proved correct.

Marx amassed a great deal of evidence about social change; what he sought was a suitable theory that would account for it. His analysis of change in society was based on the idea of a 'dialectic', a central feature of Hegel's philosophy. This involved a process in which one situation (a thesis) produces reaction (its antithesis) and the two are then resolved (in a synthesis). Marx took this theory of dialectical change and applied it to the relationships involved in the supply of society's material needs. His resulting 'dialectical materialism' saw change in terms of the conflict between social classes.

Presented in this way, Marx has clearly followed established scientific principles. He has his evidence and has framed his theory on the basis of it. However, Karl Popper claimed that Marxism was not valid science, since a Marxist would not accept that any evidence could count against the theory of dialectical materialism. If every event can be interpreted in terms of dialectical materialism, none can count against it, and it therefore fails a basic requirement that every genuine scientific theory should be theoretically capable of falsification.

However, interpretation of evidence is not the only way to assess this (or any) theory. If dialectical materialism is a valid scientific theory of social change, it should be validated or refuted by its predictions. Political upheavals in the twentieth century witnessed the power of Marxist ideology, followed by its rapid decline, since its key predictions (e.g. the collapse of capitalism) failed to be confirmed, and the process of social and political change did not appear to follow the lines of class conflict in the way that Marxist theory predicted.

SIGMUND FREUD

After training in medicine, Freud worked as a hospital doctor, taking a particular interest in neuropathology. He then set up in private medical practice, dealing with nervous conditions, especially 'hysteria'. He developed psychoanalysis as a means of exploring the origins of such conditions. Through the analysis of dreams and by the free association of thoughts, a person could be led to articulate feelings which had been locked within the unconscious, but which Freud saw as the cause of bizarre behaviour or nervous conditions.

The key question here for the philosophy of science is whether or not psychoanalysis should count as valid science, on a par with other medical sciences.

One crucial thing to keep in mind in evaluating the relevance of psychological theories to the philosophy of science is that the information provided by a person about the state of his or her mind, or the memories they describe, cannot be checked to see if it is true or false. Feelings or sensations cannot be observed directly, since they are not part of the physical world.

If someone gives an account of what he or she has experienced in childhood, one might try to question it on the grounds that it is inherently unlikely to be true, but it is very difficult to refute it. Even if it could be proved that a 'remembered' event could not possibly have taken place, that does not render the memory irrelevant – for the important thing is that, for the person undergoing analysis, the memory is significant, even if it has no foundation in fact.

With the physical sciences, it is of key importance that the experimental evidence can be reproduced in order for its accuracy to be checked. No such checking process is possible in the case of the material that forms the basis of psychoanalysis. Introspection is a valuable tool in the hands of psychologists and philosophers, but it is far from infallible, and it always has to be taken on trust.

Thus, the situation with Freud is quite different from that of Marx with respect to evidence used. For Marx it is historical, although it is possible for his interpretation to be incorrect. For Freud, it is unique and uncheckable, and it is therefore not possible to falsify the results of the analysis. It can also be argued that science requires repeatable and substantial evidence; this psychoanalysis, by its very nature, cannot provide.

Cognitive science and neuroscience

Philosophers have always debated the extent to which our knowledge comes from experience or is developed from innate ideas. But the assumption is generally made that there is a fundamental difference between knowing external things and knowing the contents of one's own mind – with the former coming through sense experience (albeit shaped by our mental operations) and the latter coming from introspection. I know my own thoughts in a way that is quite different from any other form of knowledge.

Exactly what the mind is and how it relates to the physical body is the subject matter of the philosophy of mind, and there are a whole range of options given for that relationship, with reductionist materialist views on one side and dualist views, which insist that the mind is not extended in space and is therefore quite distinct from the physical world, on the other.

By the mid twentieth century a majority of people would probably have accepted a dualist view of some sort. Over recent decades, however, the situation has changed quite radically, in part from the developments in the general area of 'cognitive science' and more recently through advances in neuroscience.

Whereas psychology had generally explored the mind in terms of the reported experience of individuals – in psychoanalysis, for example, or in the early attempts to dissect and detail every aspect of what was being experienced – the behaviourists, notably

Ivan Pavlov (1849–1936) in Russia and J. B. Watson (1878–1958) and Fred Skinner (1904–90) in the USA, approached mental operations through measuring physical responses to stimuli. Indeed, we tend to speak of a 'Pavlovian reaction' when we unconsciously respond to a stimulus; salivating, for example, at the prospect of food. The assumption of behaviourists was that a science of mind should be based on things that could be measured and quantified through experiment and observation. Many behaviourists saw mental descriptions as covert ways of describing sets of physical characteristics; if mental events were not in the physical world, they could take on verifiable meaning only by being translated into physical characteristics.

Hence being happy is measured by whether one smiles, being in pain by curling up and holding the affected part. Mind is a matter of stimulus and response. One problem with this, of course, is that people can mask their true feelings and fake others. The actor on stage is not actually feeling the emotions he or she portrays. Hence it is not straightforward to identify physical responses with experienced mental states, even if the former are generally taken (actors apart) as an indication of the latter.

Insight

We need to hold that simple, behaviourist dilemma in mind as we move forward towards the world of neuroscience. One thing may indicate another, or correspond to another, without thereby being identified with it; the experience of feeling happy is not the same thing as a statistical printout showing the number of times I have smiled.

Behaviourism was essentially a reductive approach to the human individual; it did not take into account our engagement with one another, our shared hopes and values, the way in which we learn from one another. It took a rather disengaged view of human life, whereas many would argue that you can understand human life only in its social context.

Then, in the late 1960s, there emerged a new approach to the mind, based on a combination of philosophy, psychology, linguistics and computer science. It was the attempt to approach the mind and human knowledge in a scientific way, through experiment and observation (although on a far broader basis than that undertaken by the behaviourists), rather than simply reflecting on concepts or through introspection.

16. Science and humankind

This, of course, was rather forced on philosophy by developments in parallel disciplines:

▶ **Psychology** and **pharmacology** were showing how mental processes could be influenced both through therapies and through drugs. It made no sense to say that the mind was essentially unknowable if it could be changed by drug treatment.
▶ **Linguistics** was being established as a separate discipline, using scientific methods to examine the nature of communication.
▶ **Computer science** was also starting to emerge to the point at which it made sense to ask how computers could replicate those processes, such as logic or language, that were traditionally seen as entirely non-physical, mental functions. The very question 'Can you make a computer that thinks?' breaks through the strict divide between mental and physical.

Hence **cognitive science** was born, an interdisciplinary approach to the scientific examination of human thought and knowledge. Today it is not limited to understanding the actual process of knowing something (cognition), but links with biology to examine matters related to the functioning of the brain – enter, **neuroscience**.

To put it simply, neuroscience examines the functioning of various parts of the brain, measuring the way in which the firing of neurones corresponds to, say, vision, hearing, thought and so on. The philosophical question this raises is whether such brain activity can simply be identified with the corresponding mental operation, or whether it simply describes its physical component.

Insight

In other words, a behaviourist might identify the experience of happiness with the observed smile; the neuroscientist identifies it with the firing of certain neurones. This is a gross oversimplification of what neuroscience does, but I believe the philosophical question of identification remains substantially the same.

There are huge problems with the identification of our mind – along with our hopes, fears, ambitions, thoughts and choices – with brain activity. For one thing, that activity becomes part of a physical causal series, removing personal freedom and responsibility. I may think that I am free to choose whether to steal something or not, but if it can be shown that brain activity corresponding to that choice took place just before I became aware of making it, my choosing is an illusion. I become a puppet, twitching on the end of the strings of neural activity.

This is not in any way to denigrate the scientific advances made in neuroscience and the imaging techniques that have made it possible to measure what is happening in the brain of a living subject; our ability to understand the working of the brain has grown exponentially. What it does show is the importance of examining the philosophical assumptions behind such science. Medicine benefitted greatly from seeing the human body as a machine, examining the function of each organ and seeing the ways in which they worked together to maintain life. Harvey understood that the heart was a pump, and today we are able to do considerable re-plumbing of that organ, thus preserving and enhancing life. Neuroscience has extended such knowledge into the one organ – the human brain – which had previously been inaccessible, simply because it was previously impossible to get a detailed image of what was happening in the brain while the subject was alive and thinking.

But just as my enjoyment of a run is neither diminished nor fully explained in terms of muscular action of legs and arms, raised heart rate and so on, so the mapping of the brain enhances our appreciation of the mind without thereby reducing my thoughts and emotions to the firing of neurones. And here we link back to the issue of reductionism and holism. Neuroscience – at least as it is popularly presented – seems to accept uncritically the reductionist viewpoint – that reality inheres in the action of individual or groups of neurones; most people, considering the way their life is understood in terms of their social interactions, values, lifestyle choices and so on, might find a more holistic approach to human life more suitable.

Insight

Neuroscience clearly challenges Cartesian dualism because it is able to identify and measure physical activity corresponding to mental operations, but its assumptions are generally reductionist.

Much of our mental and emotional life is generated by interactions with other people; the context and meaning of my pleasure is not simply a description of its corresponding neural activity, but is to be understood in terms of that with which I am pleased.

Neuroscience is good at describing the physical component of mental activity, but not at explaining it in terms that are meaningful to the experiencing subject.

The social function of science

To what extent should pure science be limited by the implications of technologies to which its application gives rise? No doubt there are some environmentalists who deeply regret the development of the internal combustion engine, and some scientists, including Einstein himself, were deeply concerned about the development of their research in the production of nuclear weapons. Should pure science be held responsible for its application?

From time to time a scientific theory may be put forward, based on experimental evidence, which appears to go against what is regarded as politically correct. One example of this is **Hans Eysenck** (1916–97), who, through the 1960s and 70s, did major work on intelligence testing. He came to the conclusion that there were differences in IQ related to differences in race. This led to the accusation of 'racism', implying that his findings reflected a moral and political bias. More recently, the sociobiologist O. E. Wilson's comments on racial differences while visiting London gave political offence and he was dropped from the conference at which he had been booked to speak, returning to the USA to find that he had also been sacked from his post.

Insight

This touches on a question of fundamental importance for an ethical assessment of science. Should political or moral concerns be allowed to interfere and censor the results of scientific investigation? Indeed, are there political or moral limits to what can be investigated in the first place?

Some people deliberately court controversy, and scientific findings can be presented in a way that gives offence, but what is at stake here is the authority of the scientific method: a scientist cannot claim as scientific any theory which has been produced to fit a required ideology, even if a certain amount of evidence has been discovered to support it. Rather, a theory should aim to be the best available interpretation of the available data.

There is also the issue of money. Scientific method aims to guard against the complaint that the findings of science are made to conform to the expectations of those who pay for it. It aims at objectivity, although, as will be clear by now, that is a very difficult thing to claim. At least, it is argued that science follows a methodology based on reason and evidence, rather than on political or economic expediency.

Much scientific research is paid for by industry and is conducted within defined guidelines in order to contribute knowledge that will have commercial implications. Scientists have to earn a living and therefore have to undertake work that is able to attract sponsorship or some sort of funding. This situation is not new: Archimedes was employed to develop weaponry.

Such commercial funding is not necessarily crude in its influence; it does not try to determine the results of research programmes. However, the very fact that funding is provided for those research programmes deemed worthwhile actually defines and shapes the world of research. Research which may be interesting and worthwhile, but which does not seem to be leading to any area of profitable technology, may never attract funding and therefore may never get done.

Those who engage in weapons research funded by the government, or research into food additives, paid for by a company that produces them, or into the environmental impact of various technologies, are all working to an agenda. They have been set tasks. They seek to find evidence which will be of value to those who employ them. Their work in science is driven by economic, social or political questions.

Science, however fundamental, is seldom 'pure' in the sense that it is conducted simply for the sake of increasing the sum of human knowledge. Mostly, science and technology go hand in hand, with science providing the deep theoretical work upon which technology can be built.

Example

At the time of the 'cold war' between the USSR and the USA, it was the Soviets that first succeeded in putting a satellite into orbit (*Sputnik*), and then sending a human being into space (Uri Gagarin). The response to this on the part of the US government was to increase the funding for space exploration. Partly this was for military reasons, partly also in order to show that the USA

> was unrivalled in the world of science. When President Kennedy announced the plan to land a human being on the moon before the end of the 1960s, it was a political declaration of national self-confidence.
>
> Vast amounts of resources were thrown into the development of nuclear weaponry, missile technology and guidance systems, from the 1950s, as well as the NASA space exploration programme. Hence the 'space race' was not so much a scientific as a political phenomenon.
>
> Today most of the hardware that is put into orbit is there to make a profit – mainly in the shape of communications satellites. This is not to deny that the early space programme yielded much valuable information, but simply to point out that the information gained by such research programmes and technologies was of secondary consideration when it came to funding. The pressure to succeed was political.

A valid case can be made for saying that governments should fund fundamental research, because it is from this that new ideas are born which might subsequently lead to highly profitable technologies. In this way, research can be presented as a long-term investment, with the fear that, if it is neglected, the result will be that a nation gradually slips behind in terms of science and technology, and therefore in terms of the potential benefits, both to its own citizens and to exports, that new technologies can bring.

Insight

Scientific research is socially determined. There may be a huge number of phenomena worth exploring, but only those that have some social use will be funded. Hence, even if the results of a research programme are not directly influenced by social pressures, the selection and operation of that same research programme may well be.

GIVING A DEFINITIVE ANSWER?

Experts in a particular field may be called upon to give a 'scientific' comment on a matter of public interest – for example, whether a particular food is safe to eat, a drug to take, or whether the prolonged use of mobile phones by young people can pose any threat to their health. There is a general expectation on the part of the public that science will provide a definitive answer to such questions. Yet we have seen that science always needs to remain open to the possibility (indeed, the inevitability) that our present theories will eventually be superseded. Scientific answers generally need to be qualified and interpreted carefully, but that is seldom possible in the world of media soundbites.

Here there is a real problem in terms of the clash between scientific method and the public use of scientific information. Almost every activity involves some sort of risk, but in most cases the risk is minimal and is therefore ignored. Nevertheless, in a world where people might sue for negligence, it is essential to get some professional judgement about what constitutes 'danger to health' or what is an 'acceptable risk'.

Statistics and moral choice

Statistics can predict likely outcomes, but cannot determine whether or not outcome is morally acceptable. Science can point to what is possible, but cannot say whether or not it is right.

In this context, it is worth noting the distinction between a categorical imperative (you ought to do X) and a hypothetical one (you need to do X, if you want to achieve Y). Science can provide information on which to base hypothetical imperatives – for example: 'If you want to avoid lung cancer, you should stop smoking.' What it cannot do is say that smoking is wrong.

What is clear is that all science can be expected to provide is the evidence – suitably explained – upon which a judgement can be made. It may set out the results of a series of carefully controlled tests; it may have gathered statistics from the population as a whole; it may present a host of facts and figures. What it cannot decide is what constitutes 'acceptable risk'; that is a political or moral question, not a scientific one.

In *Ethics*, the philosopher G. E. Moore pointed out that you cannot derive an 'ought' from an 'is' – a well-known error that he called the 'naturalistic fallacy'. Exactly the same fallacy applies to an attempt to define what is 'acceptable risk' or 'harmful' of 'beneficial' solely on the basis of scientific research. Qualitative judgements of this sort can be validated only with reference to the norms and values of society.

Insight

Ever since Kuhn, Lakatos and others brought into focus the way in which scientists work, there has been an interest in examining science as a social phenomenon. That perspective may undermine the 'realist' approach to scientific claims, by considering not simply what is said, but who said it and why, on the assumption that scientific claims are being backed by a financial or political agenda.

Taken to an extreme, that position could lead to cynicism about all scientific claims, which would not do justice to the genuine intention of most scientists to achieve accurate and unbiased knowledge. On the other hand, just as science needs to be careful to include all evidence in assessing a phenomenon, so the process of natural selection by political and financial backing cannot be ignored in assessing the relevance of scientific research programmes.

THE HUMAN PERSPECTIVE

The seventeenth and eighteenth centuries, during which science and technology developed and established themselves as a significant force in Western culture, were a time of some optimism. On the political as well as the scientific front there was a sense that humankind should make progress, based on reason. Research and experiment were features of science that fitted well with the overall view of life. The nineteenth century saw great developments in technology, as well as the great controversies between the science and traditional religious authority, particularly over evolution, and the groundwork for much twentieth-century science was done in the latter part of the nineteenth.

Nevertheless, science and technology did not develop totally unopposed during this period. The Romantic movement of the nineteenth century may be seen against the background of a world newly mechanized and appearing to be increasingly dominated by science and technology. Artists, poets and philosophers could make a case for a world in which human emotions, and Locke's secondary sensory qualities, took precedence over the theories of science and mathematics.

The English writer and artist William Blake (1757–1827) wasn't too happy about the technology that had sprung from science, and he was deeply opposed to the rationalism of Locke and Newton, wishing to replace it with something more imaginative and intuitive. His image of Newton is of the oppressor who measures the Earth with a pair of dividers. Hence also the 'dark, Satanic mills' that disfigured the countryside in which he seeks to build Jerusalem in his famous poem.

It was not merely the artistic temperament that railed against the increasing imposition of science and technology. From the standpoint of existential philosophy – with its emphasis on the meaning and purpose in human life – Søren Kierkegaard (1813–55) considered that science was quite capable of describing inanimate objects, plants and animals, 'But to handle the spirit of man in such a fashion is blasphemy.' (from his *Journals*, quoted by John Passmore in *Science and Its Critics* (1978)).

In many ways, the early years of the twentieth century, both in philosophy and in science, were times when progress was seen as inevitable. Science and reason had prevailed over superstition, technology had given great benefits, and all was set fair for a new world of reason and human triumph.

Then came the traumas of the twentieth century – in war; in political instability and global rivalry between conflicting ideologies; in the loss of certainty (in science, as much as in life in general). The science and technology whose benefits had remained largely unchallenged started to be questioned. The science that offered endless cheap fuel was also capable of nuclear holocaust on a global scale. Medical technology made huge strides in countering disease. Yet the demands made upon it increased faster than it could deliver. Life-saving technology was not simply marvelled at; it was demanded. A medical answer was expected for every condition. Medical and nursing ethics was born as a philosophical discipline, largely in order to cope with the human issues raised by the availability of new technology.

In the 1960s and 70s many were critical of science. The promise of a better life through technology, championed in the post-war years, appeared to many to have a downside in terms of the threat of global extinction through nuclear war, pollution of the environment, and gross inequalities fuelled by technologies and consumerism. Science was sometimes portrayed as narrow, as conducted by myopic men in white coats. It contrasted with the rising tide of self-expression, anti-war demonstrations and the hippie culture. A fascinating book from this era, looking at the criticisms lodged against science, is John Passmore's *Science and Its Critics*.

The technology that was capable of increasing crop yields through pesticides was challenged for poisoning the environment. Mechanical devices that could increase agriculture and fell forest, yielding profits from those who could now look beyond their own area, was seen as threatening a global natural resource, as rainforests diminished. The wonders of the modern aerosol and internal combustion engine were criticized for depleting the ozone layer and contributing to global warming. Global communication and information technology created wonders and also induced stress. A generation of obese children sat in front of TV sets and played computer games.

Now, in the twenty-first century, many of the human perspectives on science that were evidenced in earlier centuries remain. Technology can be a threat rather than a blessing, but its regulation is recognized as a moral and political issue rather than a scientific one. It is also astounding to reflect on just how far science and technology have come over the last decade or so.

The exponential growth in computer technology and the Internet have not only transformed and speeded up what we did before (calculating; communicating by letter or phone) but has given us possibilities not previously imagined. Cloud computing enables me to access my personal information from any point on the globe, and the whole phenomenon of social networking on the Internet opens up communication in a way that could not have been predicted a generation ago – with political as well as social consequences. I can now understand and express myself in terms of a global network which I can access in real time; I can post a comment or a photograph and have it accessed by any number of people globally. What has this done to our understanding of ourselves and the community within which we live and work? Technology, in itself, may be socially and morally neutral – but its use is certainly not.

My phone now takes photographs at a higher resolution than my camera did a few years ago, reminds me of appointments, holds all my contacts, gets me onto the Internet, sends text messages and emails, and also holds the text of every book I have written. A decade ago that would have been impossible, and yet it all so easily becomes accepted as a normal part of life. What are the implications of this for how I understand myself, how I establish and maintain friendships, and so on?

> **Insight**
>
> Aristotle argued that true friendship depends on the ability to meet and share time together. Does that still apply in the world of social media sites?

At the same time, some dramatic research projects – conducted at huge expense – are probing the fundamental features of matter and energy, and the results of experiments conducted on the Large Hadron Collider at CERN in Switzerland are global news, particularly when it appears that particles have travelled at a speed greater than that of light, and huge advances are being made in the area of genetics. There is little sense, even at a time of global economic crisis, that science should not press ahead.

However, a human perspective on science also requires that we weigh carefully exactly what science is able to do, and how its findings are justified, so that we do not accept uncritically anything that is presented as being 'scientific'. Karl Popper argued that the difference between science and pseudoscience was that the former delivered claims that were capable of being falsified. With qualifications that we have already discussed, that remains an important feature of science. The mark of true science is a commitment to truth along with a recognition of the provisional nature of every claim we make, however convincing it may seem to us at this time.

KEEP IN MIND...

1 Medicine has made progress by seeing the human body as a machine.

2 Humankind is a very recent arrival on Earth.

3 Human origins and development are inferred from very limited archaeological evidence.

4 Herbert Spencer presented 'Social Darwinism' based on the struggle for survival.

5 Karl Marx presented 'dialectical materialism' as an explanation of human social change.

6 Claims made under Freudian analysis are interpreted, but not presented as checkable evidence.

7 Behaviourism identified mental states with measurable physical responses.

8 Neuroscience tends to take a reductionist view of the mind.

9 Scientific research programmes are often subject to financial or political influence.

10 Scientific claims generally need to guard against the assumption of absolutism.

Postscript

Some have seen the function of the philosophy of science as limited to an exploration of the methods by which science operates, the language it uses, and the way in which it justifies its claims. That is, of course, a perfectly valid function, but it leaves out several crucial questions. Just as one cannot imagine studying ethics without taking into account the impact on society of the actual moral choices that people make, so it seems inadequate to have a philosophy of science that does not also address the impact that its knowledge and (through technology) its control over nature has had.

It is clear that many of those who laid the foundations of modern science were optimistic about the benefits to be gained from the application of human reason. They were of the view that reason would replace crude superstition, that life could and should be examined, and that the world could be understood in a way that would allow humankind a more positive measure of control over its destiny. This very positive goal of natural philosophy was found in thinkers such as Francis Bacon and René Descartes. Indeed, Descartes made it clear in his *Discourse on Method* (1637) that he was deliberately moving away from speculative philosophy towards a practical philosophy that would aid humankind through the control of nature.

And, of course, the human world has been profoundly changed by the developments of science and technology. Few people today would wish to reject all the benefits of modern medicine, communication or travel. In this sense, it is difficult to argue with the claim that science has been humankind's biggest success. Species have generally survived by adapting to their environment; humankind has gone a long way to adapting its environment to suit its own survival.

The process of scientific investigation is one of abstraction, of framing principles that enable predictions to be made. The phenomena considered by science are therefore quite different from the sensual experiences enjoyed by human beings. If human experience is multicoloured, then some have seen in science the tendency to reduce everything to a mathematical grey. This was the complaint of romantics like William Blake, and is implied by modern complaints about the 'reductionism' of science.

We therefore need to remember that science is one, but only one, of the ways of encountering and understanding reality. If I am to fall in love, have a religious experience, act morally, or produce something creative in the arts, I need skills quite other than those of reason and analysis. Looking, responding, ascribing value: these are human ways of relating to the world that are non-scientific.

Science is therefore a valuable but limited way of encountering the world. In reality, it has never claimed to be anything else. There are other ways of seeing.

However, this is far from saying that science presents a mundane, factual, unexciting view of the world, to be contrasted with the sense of awe generated by the arts or by religion. Richard Dawkins, in his book *Unweaving the Rainbow* (1998), argues that the amazing features of our world, revealed by science, serve only to promote a sense of awe and wonder at nature. He shows that there is something amazing in the fact that we are alive at all, that the Earth provides exactly what is needed for human life to have evolved. A brief reflection shows that so many things have had to have happened in the past in order for us to have been conceived and born in this time and place. A minor chance occurrence to one of our ancestors, and we would not exist in the way we do. In a sense, he shows how fragile and unlikely our life is, but also how wonderful. Facing the facts about how the most beautiful things have come about – unweaving the rainbow, and seeking to understand the science of light's different wavelengths – is not to decrease their power to induce wonder, but to enhance it. He regards the whole effort to understand the world in which we live for such a brief time, as in itself a noble quest.

Science offers a positive heuristic; in other words, it seeks to find answers to problems, largely on the basis of trial and error, in an ongoing programme of research. In this, it is fuelled by the need both to create practical technologies to solve various human problems and to satisfy human curiosity by finding answers to fundamental questions.

In looking at the philosophy of science, we tend to think that pure science – the desire to know, freed from the need to show practical use – is fundamental, and that technology comes along afterwards, bringing with it both benefits and threats. In fact, it was only the availability of a leisured class (in ancient Greece, or eighteenth-century England), able to spend time in general speculation, that has produced the phenomenon of pure science or pure philosophy; for most of the time human beings have been thinking and creating, their work has been in response to need. It has been demand led, and it has justified itself by its results. Medical science does not develop in a completely healthy world!

We have already noted that, where funding for science is given for commercial reasons and new technologies are developed in order to be exploited in the global marketplace, the way in which scientific research programmes are evaluated is likely to be quite different from any theoretical situation in which scientists are unaffected in their deliberations by the necessities of trade or the approval of the *populus*. A broadly based philosophy of science, therefore needs to recognize and monitor what we may call the social or participatory aspects of science, for science has developed in response to human inquisitiveness and human need, and its success is generally judged by its ability to solve problems that each present.

Science does not, and cannot, answer all human questions. In particular, as we saw earlier, it cannot provide a value-based justification for its own activity. Science says what is possible; it is then left up to those concerned with the law and with ethics to decide whether what is possible should be made actual. The astounding contribution of modern science is that it has given humankind so many possibilities, which can be used for good or ill.

Philosophy is the 'love of wisdom'; a 'philosophy of science', in the broadest sense, is clearly essential if humankind is to benefit from what science continues to achieve.

Further reading

There are many substantial introductions to the philosophy of science for the serious student, and a huge number of books covering particular issues. Those listed here are no more than a personal selection for your next steps.

A. F. Chalmers, *What is this thing called science?*, Open University Press, 3rd edn, 1999. This takes a very readable approach.

James Ladyman, *Understanding Philosophy of Science*, Routledge, 2002. This is a very clearly written book and does not assume knowledge of philosophy.

Marc Lange (ed.), *Philosophy of Science*, Blackwell, 2007. There are basic introductions to each section of this book, but there are some serious and semi-technical papers here, and the text can be quite hard-going at times.

Anthony O'Hear (ed.), *Philosophy of Science*, Cambridge University Press, 2007 (the Royal Institute of Philosophy Lectures 2005/06).

Alex Rosenberg, *Philosophy of Science*, Routledge, 2000.

For an anthology covering many of the key articles and central questions, see:

Martin Curd and J. A. Cover, *Philosophy of Science: The Central Issues*, Norton & Co., 1998.

For the serious student, David Papineau (ed.), *The Philosophy of Science*, Oxford Readings in Philosophy, Oxford University Press, 1996, is an important collection of key themes, which are readable although densely packed.

The Philosophical Papers of Imre Lakatos, published by Cambridge University Press in 1978, four years after his death, are a valuable collection of his work on the philosophy of science, summarizing some of the key debates of the mid twentieth century with clarity. The first volume, *The Methodology of Scientific Research Programmes*, is particularly useful.

Particular areas within the philosophy of science are explored in readable fashion in the following:

For an historical and philosophical introduction to the issues of scientific method: Barry Gower, *Scientific Method*, Routledge, 1997.

Jack Ritchie, *Understanding Naturalism*, Acumen, 2008, touches on many other aspects of the philosophy of science as well.

Ian Hacking, *The Taming of Chance*, Cambridge University Press, 1990, looks at the way in which the gathering of statistics led social scientists to frame 'laws' that predicted outcomes without taking away the freedom of individuals to select how they should act.

For the whole issue of induction, Nelson Goodman, *Fact, Fiction and Forecast*, 4th edn, Harvard University Press, 1983, is the classic text, originally appearing in 1954 and still much quoted.

For a modern account of evolution, try Steve Jones, *Almost like a Whale*, Doubleday, 1999.

For issues relating to the human sciences, see: Charles Taylor, *Philosophy and the Human Sciences*, Cambridge University Press, 1996.

For a fascinating collection of articles on science – not strictly on the philosophy of science, but raising many issues and whetting the appetite to ask questions, see: John Gribbin, *The Case of the Missing Neutrinos*, Penguin, 2000; and covering far more issues than even the title would suggest: Jack Cohen and Ian Stewart, *The Collapse of Chaos: Discovering Simplicity in a Complex World*, Penguin, 2000.

For cosmology, try: Barbara Ryden, *Introduction to Cosmology*, Addison Wesley, 2003; or the readable and illustrated Stephen Hawking, *The Universe in a Nutshell*, Bantam Press, 2001.

For the history of the debates over quantum theory: Andrew Whitaker, *Einstein, Bohr and the Quantum Dilemma*, Cambridge University Press, 2006.

Richard Dawkins' books provide a wonderfully readable overview of science: *The Blind Watchmaker* (1986) gives a fascinating account of evolution by natural selection; *Climbing Mount Improbable* (1996) shows how the complexity of life can be accounted for through the small incremental changes brought about by evolution, and *Unweaving the Rainbow* (1998) shows how a scientific analysis, far from detracting from a sense of wonder, actually expands and deepens it. (All available from Penguin.)

Karl Popper's Philosophy of Science: Rationality without Foundations by Stefano Gattei, Routledge, 2009, is a short, accessible book, giving a wonderfully clear and enthusiastic exposition of Popper's philosophy.

For an examination of the personal and religious issues raised by science, see *Religion and Science* in Hodder & Stoughton's Access to Philosophy series.

> For further information about these and other books and for a list of Internet resources on the philosophy of science, visit the author's website: www.philosophyandethics.com

Glossary

analytic statements Those whose truth is established by definition (e.g. logic and mathematical statements), rather than by evidence (see **synthetic statements**).

atomism The theory (first put forward in the fifth century BCE) that all matter is composed of atoms separated by empty space.

correspondence theory The theory that the meaning of a word is given by the object to which that word corresponds (problematic if we have no independent knowledge of objects).

determinism The philosophical view that all things are totally conditioned by antecedent causes.

epicycle The path traced be a point on the circumference of a circle as that circle is rolled around the circumference of a larger one; used for calculating the orbits of planets up to the seventeenth century.

falsification Used by Popper as a criterion for genuine scientific claims, namely that they should be capable of being falsified on the basis of contrary evidence.

the final cause Purpose of something; the actualization of its essence and potential (in Aristotelian philosophy).

foundationalism The attempt to establish a basis for knowledge that requires no further justification (i.e. a point of certainty).

holistic Describes an approach, argument or view that considers the operations of the whole of a complex entity (as opposed to its constituent parts).

hypothetico-deductive method The process of devising and testing out explanatory hypotheses against evidence.

induction (inductive inference) The logical process by which a theory is devised on the basis of cumulative evidence.

instrumentalism The view that scientific theories are to be assessed by how effective they are at explaining and predicting the phenomena in question.

light year The distance travelled by light in one year, at a speed of 186,000 miles per second.

Logical Positivism A school of philosophy from the first half of the twentieth century, which, influenced by the success of science, attempted to equate the meaning of a statement with its method of verification.

materialism The view that the world consists entirely of physical material (hence opposing any form of mind/body dualism or supernaturalism).

natural philosophy The branch of philosophy which considers the physical world, a term used to include science prior to the eighteenth century.

Ockham's Razor The principle that one should opt for the simplest explanation; generally summarized as 'causes should not be multiplied beyond necessity'.

paradigm A theory or complex of theories which together set the parameters of what is accepted as scientifically valid within its particular sphere of study. Kuhn describes how paradigms may eventually be replaced if they prove inadequate.

phenomena Those things which are known through the senses; in Kant, it is the general term used for sense impressions, as opposed to **noumena**, or things as they are in themselves.

primary qualities A term used by Locke for those qualities thought to inhere in objects, and are therefore independent of the faculties of the observer (e.g. shape).

realism The view that scientific theories are descriptions of independent and actual, if unobservable, phenomena.

reductionist Used of a process which analyses complex entities into their component parts, and (by implication) ascribes reality primarily to the latter.

relativism The view that all statements about the world are made relative to some particular viewpoint and thus cannot be objective or final.

scientism The view that science gives the only valid interpretation of reality.

secondary qualities A term used by Locke for those qualities used in the description of an object that are determined by the sensory organs of the perceiver (e.g. colour).

spacetime singularity A theoretical point of infinite density and no extension, from which the present universe, including space and time themselves, is thought to have evolved.

synthetic statements Those whose truth depends upon evidence (see **analytic statements**).

TOEA 'theory of everything', the attempt to find a single theory to account for the four fundamental forces (or interactive forces) of nature (gravity; electromagnetic; strong and weak nuclear) which describe the way in which elementary particles interact.

underdeterminism The situation in which we do not have enough evidential data to choose between competing theses.

utilitarianism Theory by which an action is judged according to its expected results.

Weltanschauung Term used for an overall view of the world, through which experience is interpreted.

Index